Momentum Wave Functions-1976
(Indiana University)

AIP Conference Proceedings
Series Editor: Hugh C. Wolfe
Number 36

Momentum Wave Functions-1976

(Indiana University)

Editor

D. W. Devins

(Indiana University)

American Institute of Physics
New York   1977

Copyright © 1977 American Institute of Physics, Inc.
This book, or parts thereof, may not be reproduced
in any form without permission.

L.C. Catalog Card No. 77-821-45
ISBN 0-88318-135-5
ERDA CONF- 760567

American Institute of Physics
335 East 45th Street
New York, N.Y. 10017

Printed in the United States of America

Foreword

The Workshop/Seminar on Momentum Wave Function Determination in Atomic, Molecular and Nuclear Systems was held in Bloomington, Indiana, 31 May - 4 June, 1976.

This workshop/seminar was held under the auspices of the U.S.-Australian Scientific Cooperation Program administered by the National Science Foundation. The publication costs were underwritten jointly by the Office of the Dean of Graduate Research and Development of Indiana University and the Indiana University Cyclotron Facility.

The workshop coordinators were
R.A. Bonham, Professor of Chemistry
D.W. Devins, Professor of Physics
G.E. Walker, Professor of Physics

D.W.D.

TABLE OF CONTENTS

Momentum Determinations by Knockout Reactions
    I.E. McCarthy 1

Intermediate Energy Nuclear Reactions
    G.B. Walker 21

(p,2p) Experiments at the University of Maryland Cyclotron
    P.G. Roos 32

Investigation of Atomic and Molecular Electronic Structure
by Use of the (e,e') Reaction
    R.A. Bonham 51

The IU-Melbourne (p,2p) Experiments at IUCF
    D.W. Devins 77

The Determination of Electronic Momentum Distribution and
Atomic and Molecular Structure using the (e,2e) Reaction
    Erich Weigold 84

Nucleon Knockout: Reaction Mechanisms
    Edward F. Redish 111

Quasi-Free (p,pα) Scattering at 157 MeV
    P. Radvanyi 127

Green's Function Method to the ($\hbar\omega$,2e)-Reactions
    P. Winkler 128

Theoretical Studies of the Momentum Distribution of
Molecular Hydrogen
    Vedene H. Smith, Jr. 145

Investigation of (e,2e) Knockout Reactions via Electronic
Structure Calculations
    Geoffrey R.J. Williams 151

Nuclear Physics in Australia
    B.M. Spicer 168

Nucleon Knockout: Offshell Effects
    G.J. Stephenson, Jr. 169

High Resolution (p,2p) Studies at 800 MeV
    Robert K. Cole 174

Quasi-Free ($\pi, \pi N$) Scattering
    V.E. Herscovitz, Th. A.J. Maris, P.M. Mors,
    C. Schneider 179

A Knockout Reaction Study with a Polarized Beam
    P. Kitching. C.A. Miller, D.A. Hutcheon, A.N. James,
    W.J. McDonald, J.M. Cameron, W.C. Olsen, G. Roy 182

Three-Body Collisions Involving Breakup
    Ian H. Sloan 187

On the (e,2e) Reaction in Solids
    N.R. Avery 195

Validity of the (e,2e) Reactions as a Probe of the Atomic
and Nuclear Structure
    A. Giardini-Guidoni, R. Tiribelli, D. Vinciguerra,
    R. Camilloni, G. Stefani, G. Missoni 205

Distorted Wave Calculations for (p,2p) Reactions
    R.D. Koshel 227

Recent H.E.E.I.S. Results on the Compton Defect
    A.D. Barlas, W. Rueckner, H.F. Wellenstein 241

Born Approximation Calculations of the Compton Defect in
Aluminum
    L.B. Mendelsohn, H. Grossman 249

The Compton Defect: Is There a Shift in the Compton Peak?
    I.E. McCarthy, R.A. Bonham 255

Special Aspects of the Nuclear Problem
    R.M. Thaler 263

Workshop Summary
    R.A. Bonham 272

Workshop Participants 277

MOMENTUM DETERMINATIONS BY KNOCKOUT REACTIONS

Lecture at the 1976 Workshop on Momentum Wave Functions,
Bloomington, Indiana

I.E. McCarthy

Institute for Atomic Studies
School of Physical Studies
The Flinders University of South Australia
Bedford Park, S.A. 5042, Australia

A knockout reaction is observed by a kinematically complete experiment in which two particles A and B emerge from the interaction region, one particle A being identical to the incident particle 0. The momenta of both are measured in coincidence. The separation energy ε of B and the recoil momentum $\underline{q}$ of the residual (core) system are both determined in the experiment.

$$\varepsilon = E_0 - E_A - E_B ,$$
$$\underline{q} = \underline{k}_0 - \underline{k}_A - \underline{k}_B . \tag{1}$$

For reasonably high incident energies the reaction mechanism may be understood as the direct knockout of B by A. The core-particle interactions distort the ingoing and outgoing wave functions from the plane wave form. If distortion is small the core acts as a spectator in the reaction and the recoil momentum $\underline{q}$ is equal and opposite to the momentum of B at the collision instant.

At the start, since the workshop concerns both nuclear and atomic momentum measurements, we should observe the close phase-space correspondence between the two domains so that both atomic and nuclear physicists can develop a feeling for orders of magnitude in the other field.

The nuclear unit of momentum is the inverse fermi (fm^{-1}). The energy of a proton whose momentum is $1 fm^{-1}$ is 20.75 MeV. The atomic unit (a.u.) of momentum is the inverse Bohr radius of hydrogen (a_0^{-1}). The energy of an electron whose momentum is 1 a.u. is 13.6eV. Therefore we can think of 1fm and a_0 as having parallel space significance and 1 MeV and 1eV as being the corresponding energy parallel.

The first knockout experiment was a (p,2p) experiment performed by Chamberlain and Segré[1] in 1952 using 340 MeV protons on lithium. The results are shown in fig. 1. Here the angle ϕ of one proton detector was fixed at 45° and the angle ψ of the other detector varied. The separation energy ε was not resolved. The angular correlation clearly shows a distribution of recoil

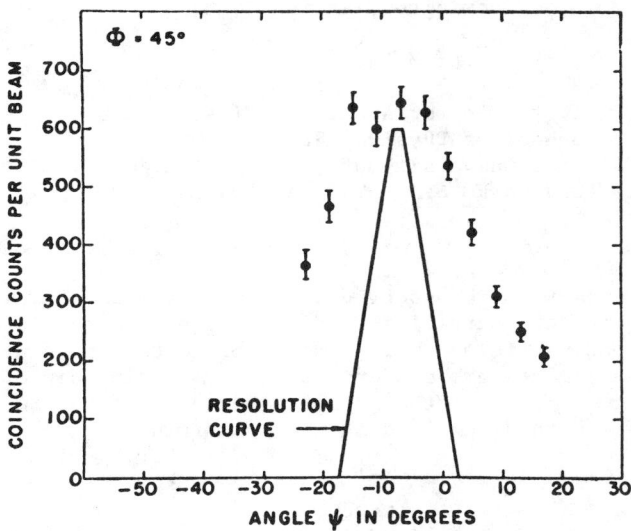

Fig. 1. The total number of coincidence counts for variation of the angle ψ of one detector with the other set at 45° in the 340 MeV (p,2p) experiment on lithium of Chamberlain and Segrè. Reproduced from reference 1.

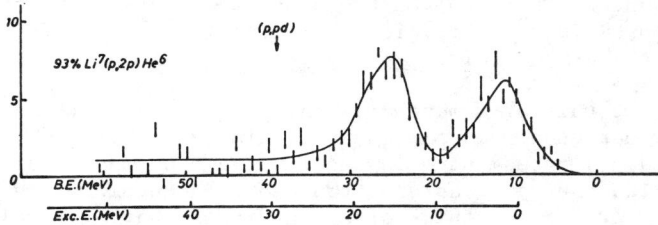

Fig. 2. The number of coincidence counts in the 180 MeV ^{7}Li (p,2p)^{6}He experiment of Tyrén, Hillman and Maris plotted against separation energy for fixed angles. Reproduced from reference 3.

momentum q about the quasi-free (q=0) maximum, which is at about
$\phi + \psi = 90°$. The distribution can be fitted by assuming the
knockout mechanism and that the protons in Li consitute a
degenerate Fermi gas[2] cut off at $q = 1\text{fm}^{-1}$. This is clearly
consistent with independent-particle model ideas of nuclear
structure. It constitutes the first use of a knockout reaction
to yield information about bound state momenta. It also confirms
the knockout mechanism.

The nuclear physics world was really set alight by the 1958
Uppsala (p,2p) experiments of Tyrén, Hillman and Maris[3]. An
example, ^{7}Li is shown in fig. 2. Here the differential cross section is plotted against the separation energy ϵ. The figure is a
dramatic confirmation of the shell model. A peak occurs at the
energy expected for $1p_{3/2}$ level and another shows where the 1s
level is.

Fig. 3 shows the first published experiment, again by Hillman, Tyrén and Maris[4] for ^{7}Li, in which recoil momentum distributions (angular correlations) were determined for individual
single-particle orbitals. Here the 1s distributions peaks at q=0
and the 1p distribution has a minimum at q=0, since s particles
have a high probability of having zero momentum and p particles
have angular momentum, which means that they are unlikely to have
zero momentum.

The angular correlations, under the assumption of no distortion, reflect directly the momentum-space orbitals. The experiment is a measurement of all there is to know about single-particle orbitals.

Part of the measuring process is of course the theoretical
interpretation of the data. We would like to find a theory that
fits the data in detail so that we have an accurate and sensitive probe for the structure information.

Early interpretations used the plane-wave impulse approximation for the knockout cross section:

$$\frac{d^3\sigma}{d\Omega_A d\Omega_B dE_A} = K \left(\frac{d\sigma}{d\Omega}\right)_{fr} \Sigma_m |M'(q,\epsilon)|^2 , \quad (2)$$

$$M'(q,\epsilon) = <\underline{k}_A|<\underline{k}_B|\psi_i(\epsilon)>|\underline{k}_0> \quad (3)$$

$$= <\underline{q}\,|\psi_i(\epsilon)> . \quad (4)$$

Here we have included the effects of antisymmetrization. The
orbital wave function for the set i of quantum numbers is $\psi_i(\epsilon)$.
The set i includes the projection quantum number m, for different

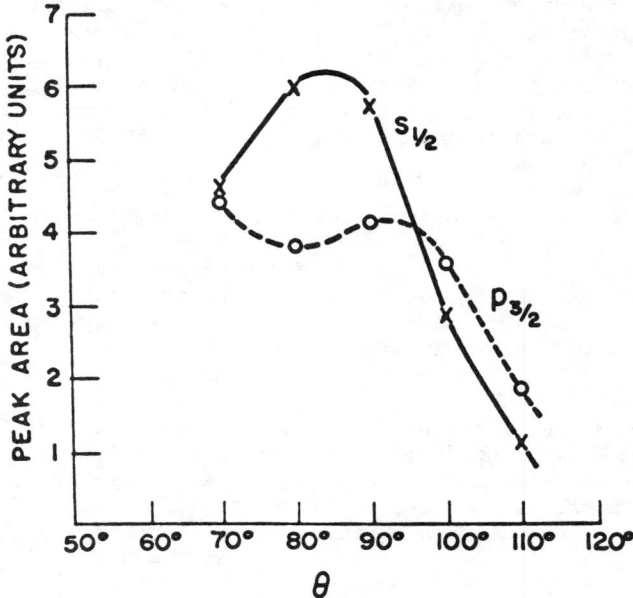

Fig. 3. The number of coincidence counts for resolved single-particle orbitals in the 180 MeV ^{7}Li(p,2p)^{6}He experiment of Hillman, Tyrén and Maris plotted against the total angle θ subtended by the counters in coplanar symmetric geometry. Reproduced from reference 4.

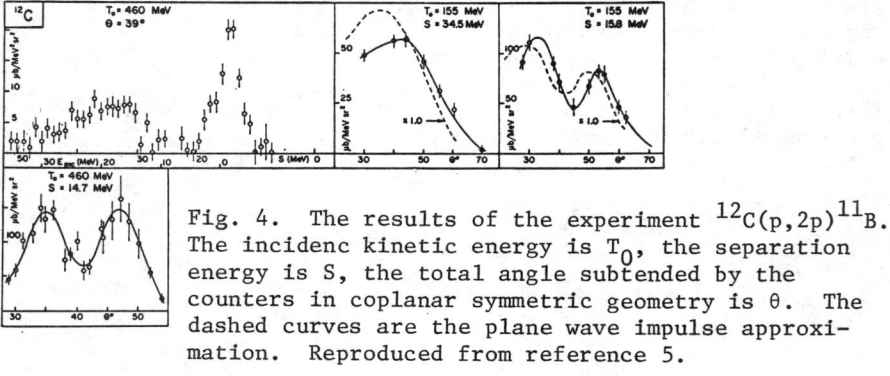

Fig. 4. The results of the experiment ^{12}C(p,2p)^{11}B. The incidenc kinetic energy is T_0, the separation energy is S, the total angle subtended by the counters in coplanar symmetric geometry is θ. The dashed curves are the plane wave impulse approximation. Reproduced from reference 5.

values of which the orbital energy ε is degenerate. There is a kinematic factor K and a factor $(d\sigma/d\Omega)_{fr}$, which is the cross section for a collision of the two particles in the absence of the core, calculated at some appropriate energy and scattering angle. The fact that there is no unique prescription for this energy is one of the disadvantages of the method, but it is not a big disadvantage for high energy (p,2p) where $(d\sigma/d\Omega)_{fr}$ varies slowly.

Fig. 4 is fairly typical of high energy (p,2p) data, in this case 460 MeV and 155 MeV for ^{12}C. The figure was taken from the review article of Jacob and Maris[5]. The 460 MeV data are due to Tyrén, Kullander, Sundberg, Ramachandran and Isaacson[5], and the 155 MeV data are due to Garron, Jacmart, Riou, Ruhla, Teillac and Strauch[6]. They use coplanar symmetric geometry: $\phi_A = 0$, $\phi_B = \pi$, $\theta_A = \theta_B = 0$, $E_A = E_B = 1/2E$. The dotted line is an impulse approximation calculation by Sakamoto[7]. The calculation reproduces the qualitative features of the data, but not details of peak positions, absolute magnitude or peak height ratio.

I want to consider how one would go about deriving a theoretical expression for the knockout amplitude M in the case of two equal-mass particles. It has been known for a long time that nucleon-nucleus collisions are dominated by the elastic channel, in the sense that the elastic cross section is usually much greater than the cross section for any one nonelastic channel. Under these conditions the proton-core interaction may be described by an energy-dependent optical model potential, which depends only weakly on particle number, os that the entrance channel potential is approximately the proton-core potential for laboratory energy E_0.

It is possible to include core degrees of freedom explicitly in the derivation, but the success of the optical model already makes the ensuing result plausible, namely that we may consider the knockout problem as a three-body problem governed by the proton-core potentials $V_A(\underline{r}_A)$ and $V_B(\underline{r}_B)$ and the proton-proton potential $v(r)$.

We take the core mass to be infinite for simplicity. The coordinates and momenta may now be written either in terms of the particle motion or of the center-of-mass and relative motion.

$$\underline{r} = \underline{r}_A - \underline{r}_B \, , \; \underline{k}' = \tfrac{1}{2}(\underline{k}_A - \underline{k}_B), \; \underline{k} = \tfrac{1}{2}(\underline{k}_0 + \underline{q}),$$

$$\underline{R} = \tfrac{1}{2}(\underline{r}_A + \underline{r}_B) \, , \; \underline{K}' = \underline{k}_A + \underline{k}_B, \; \underline{K}' = \underline{k}_0 - \underline{q} \, . \qquad (5)$$

In the quasi-three-body approximation the amplitude is exactly written in terms of a distorted-wave representation as follows

$$M = A <\chi_A^{(-)}(\underline{k}_A)| <\chi_B^{(-)}(\underline{k}_B)| (f|v$$

$$+ v \frac{1}{E^{(\pm)} - (K_A + K_B + V_A + V_B + v)} v|g> |\chi_0^{(+)}(\underline{k}_0)>, \quad (6)$$

where A is the antisymmetrization operator, $|f)$ is the observed final core state, $|g>$ is the target ground state,

$$K_A + K_B = K_r + K_R \qquad (7)$$

is the operator for the total kinetic energy, and the distorted waves $\chi_J^{(\pm)}$, $J = 0$, A or B, are particle-core scattering eigenfuctions

$$\left[E_J^{(\pm)} - K_J - V_J\right] \chi_J^{(\pm)} = 0. \qquad (8)$$

To simplify (6) we notice that it would become extremely simple if we could commute the final-state distorted waves with $v(r)$ so that the resolvent would operate directly on its eigenstate. This is done approximately by making a Taylor expansion of each potential, V_A, V_B about $\underline{R}$ and neglecting gradients with respect to $\underline{r}$. In fact gradients of optical model potentials are not large and this should be a good approximation. It becomes

$$M = <\chi_A^{(-)}| <\chi_B^{(-)}|T(p^2)|(f|g>|\chi_0^{(+)}>, \qquad (9)$$

where

$$p^2 = \hbar^2 k'^2/m \qquad (10)$$

and T is the antisymmetrized particle-particle T-matrix. Note that in this approximation the core eigenstate $|f)$, which does not depend on the particle coordinates, may be commuted through T so that the structure enters the expression only in the form of the core-target overlap $(f|g>$.

Approximation (9) is called the distorted-wave off-shell impulse approximation. Unfortunately it cannot be computed. We simplify it still further for high particle energies by considering the form of the distorted waves $\chi_J^{(\pm)}$.

They are represented to a good approximation by attenuated plane waves $\exp(i\underline{\kappa}_J^{(\pm)} \cdot r)$, where

$$\underline{\kappa}_J^{(\pm)} = (1 + \beta_J \pm i\gamma_J)\underline{k}_J, \qquad (11)$$

the term β_J being included to take acount of minor changes in the effective wave number.

In this approximation, and neglecting the small parameters β_J, γ_J in the arguments of the T-matrix, (9) becomes

$$M = <\underline{k}'|T(p^2)|\underline{k}> <\chi_A^{(-)}(\underline{k}_A)|<\chi_B^{(-)}(\underline{k}_B)|(f|g>|\chi_0^{(+)}(\underline{k}_0)>. \quad (12)$$

We may retain the averaged eikonal approximation (11) only in the first factor, restoring the fully-distorted waves in the second factor. Approximation (12) is now the factorized distorted-wave off-shell impulse approximation. It is very like the simple approximation (2), (3) except that the (off-shell) kinematics of the T-matrix is well-defined, distorted waves are used, and we see how, if necessary, to use a more sophisticated structure model than the single-orbital model.

The second factor in (12) was calculated in 1964 by Lim and McCarthy[8]. Distorted waves make no serious qualitative differences to matrix elements. In particular they worsen if anything the discrepancy in the peak height ratios for which, in the ^{12}C example of Fig. 4, the theoretical ratio of the left to right peak is too large.

The present status of the use of impulse approximations for high energy (p,2p) is that the peak height ratio is probably the worst qualitative discrepancy. Some hope that this may be removed within the framework of the off-shell impulse approximation is given by Fig. 5, which shows the squared off-shell T-matrix as a function of θ using the kinematics of the 150 MeV (p,2p) reaction on the $1p_{3/2}$ orbital of ^{12}C. The T-matrix has been computed by Birrell, McCarthy and Noble[9] for various realistic pp potentials, some of which give a reduction of the left-right ratio for the peaks, which are at about 30° and 60°. Given a thorough understanding of the reaction mechanism, we could regard (p,2p) as a means of distinguishing nuclear forces.

One must ask whether (12) is a good approximation to the three-body problem. Some information on this question is available from a calculation by Young and Redish[10], in which the Faddeev and impulse solutions to a three body separable-potential model, which represents $^4He(p,2p)^3H$ quite well, are compared. The approximation (12) is good at 100 MeV, but not at 65 MeV.

One must also ask whether the quasi-three body mechanism is a correct description of the physics. There is fairly strong evidence now that many-body resonances play a decisive part in (p,2p) experiments in the 50 MeV range. Geramb and Eppel[11] have used a giant resonance description with qualitative success. One of the interesting experiments for the IUCF will be to investigate the probable disappearance of resonance ("two step") effects as the

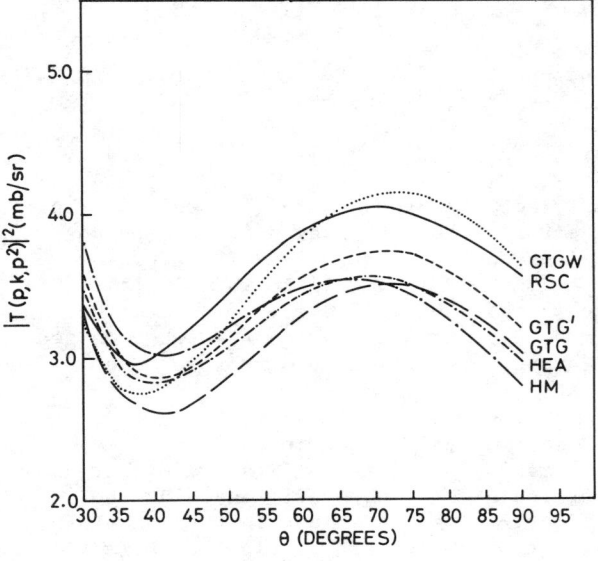

Fig. 5. The squared off-shell pp T-matrix for 150 MeV $^{12}C(p,2p)^{11}B$ as a function of the coplanar symmetric half-angle θ for various realistic nuclear forces. The forces are identified in reference 9.

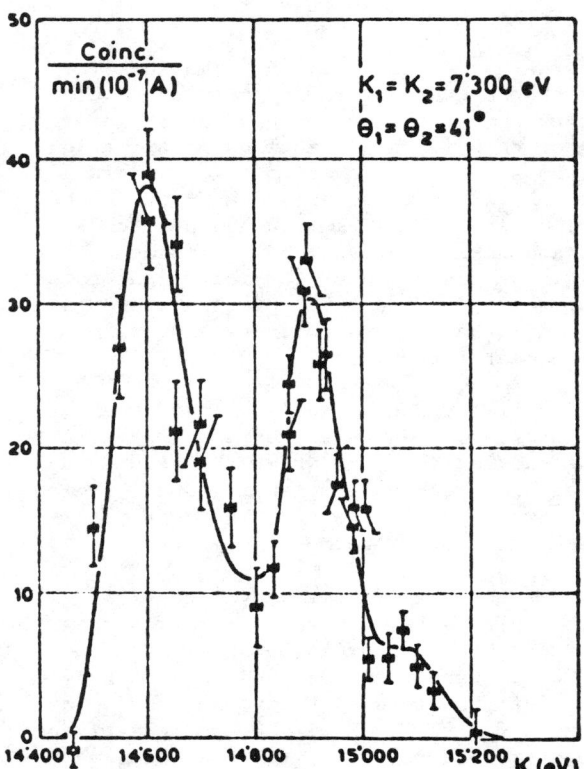

Fig. 6. Number of coincidence counts for variation of the incident energy with fixed final-state conditions in the (e,2e) experiment of Amaldi *et al* on carbon. Reproduced from reference 13.

incident energy increases.

An unfortunate aspect of (p,2p) experiments in the past has been that actual eigenstates of the residual nucleus could not be resolved in the higher-energy experiments where the three-body mechanism can be trusted. The IUCF provides hope that (p,2p) will soon be used as a tool for nuclear structure information with more detail than just single-particle orbitals. It is superior to other methods such as (p,d) because of its sensitivity to the shape of the orbitals. It makes possible the understanding of structure in terms of the best possible independent-particle basis.

From the foregoing discussion we have seen that the (p,2p) experiment promises to be a decisive tool for investigating nuclear structure, and possibly nuclear forces. The fact that it has not yet lived up to its expectations has been due largely to experimental difficulties, which are about to be overcome. There is also some remaining doubt as to the validity of the best theory we have for the description, the factorized distorted-wave off-shell impulse approximation.

I now want to turn to the analogous experiment in the atomic and molecular domain, (e,2e). For the nuclear physicists it provides an analogue computer, in which we can test the reaction theory. If we have a good reaction theory, and good enough experiments, we can use it as a probe for almost all the structure information, in the way we would like to use it in nuclear physics.

The potential value of the (e,2e) experiment was recognized in the literature[12] as soon as the (p,2p) reaction had demonstrated the value of knockout reactions. However, it was not until the experiment of Amaldi, Egidi, Marconero and Pizzella[13] that its experimental feasibility was demonstrated. Fig. 6 shows the variation of incident energy in a coplanar symmetric experiment with fixed final state conditions corresponding to $q \sim 3$ a.u. The 1s-shell electron peak has a separation energy about 300 eV higher than the valence peak. The carbon 1s energy is 283 eV.

The first (e,2e) experiment to give a recoil momentum profile for a resolved single-particle state was published in 1972 by Camilloni, Giardini-Guidoni, Tiribelli and Stefani[14]. The state was the 1s state of solid carbon. Fig. 7 shows the angular correlation in a coplanar symmetric experiment and also the counting rate as a function of the recoil momentum q in keV/c (1 a.u. is ~ 4 eV/c). The sensitivity to the shape of the orbital wave function is clearly demonstrated. The Roothaan wave function gives a much lower χ^2 than a minimum basis set function.

In the 1973 experiment of Weigold, Hood and Teubner[15] on argon, valence states were resolved for the first time and the importance of electron-electron correlation effects was un-

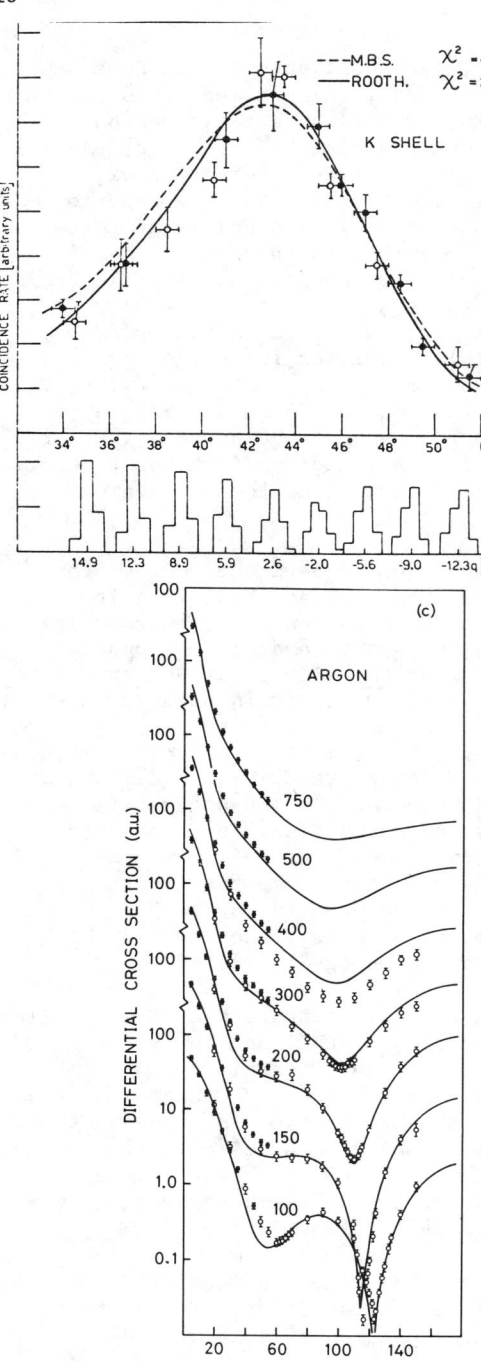

Fig. 7. The K shell momentum profile in the 9 keV Coplanar symmetric (e,2e) experiment of Camilloni *et al* on carbon. The theoretical curves are computed with the plane wave impulse approximation. Reproduced from reference 14.

Fig. 8. Differential cross sections for electron elastic scattering by argon at different energies. The theory is the optical model of McCarthy *et al*[16].

ambiguously established by observing two groups of final states, each of which could be assigned 3s symmetry by observing the q-distribution.

Since this first experiment, (e,2e) experiments have been performed on the valence electrons of atoms and molecules with energy resolutions of a few eV. Since states of the residual ion are typically separated by a few eV, these experiments have the great advantage that pure final electronic states can be investigated. For molecules final state rotational and vibrational bands built on the electronic band heads are not resolved and closure eliminates them from the theory.

Atomic and molecular physics does not have the problem of expensive accelerators. Resolution of several eV can be achieved at energies ranging from a few hundred eV to a few thousand eV, just the range in which distorted waves are accurately given by an optical model[16]. Since the optical model is less familiar in atomic physics than in nuclear physics, we show the results for elastic differential cross sections on argon at different energies in fig. 8. There are no free parameters in this model.

For energies of a few thousand eV plane waves are a sufficient approximation for valence electron studies. This can be understood from fig. 9 which shows the wave fronts for 200 eV electrons on the xenon ion. The distorted wave is not very different from a plane wave (possibly with modified wave number) in the valence region Distortion at this energy is very marked near the nucleus. As the energy increases, distortion becomes less significant.

At Flinders University two geometries have been used for the experiment. Coplanar symmetric geometry varies the energy p^2 of the Mott scattering T-matrix in equation (12) as rapidly as possible, thus changing the kinematic conditions of the experiment drastically and providing a strict test of the theory.

Figures 10 and 11 show that the theory of equation (12) survives the test admirably at 400 eV for argon (where distortion is very strong). Plane waves (dashed lines) are inadequate in this situation. However for the 3p orbital of argon at 1000 eV (fig. 12) plane waves cause an error of only $1°$ at the low-angle peak and no error for the high angle peak. The plane wave approximation is certainly reasonable above 1000 eV and it is used for molecules, where distortion due to first-row atoms is less.

These calculations were done by Noble[18]. They give us reason to believe that the off-shell impulse approximation is a sound approximation to quasi-three-body knockout and that it should be adequate for analyzing future (p,2p) data.

In noncoplanar symmetric geometry, the angles θ are fixed at

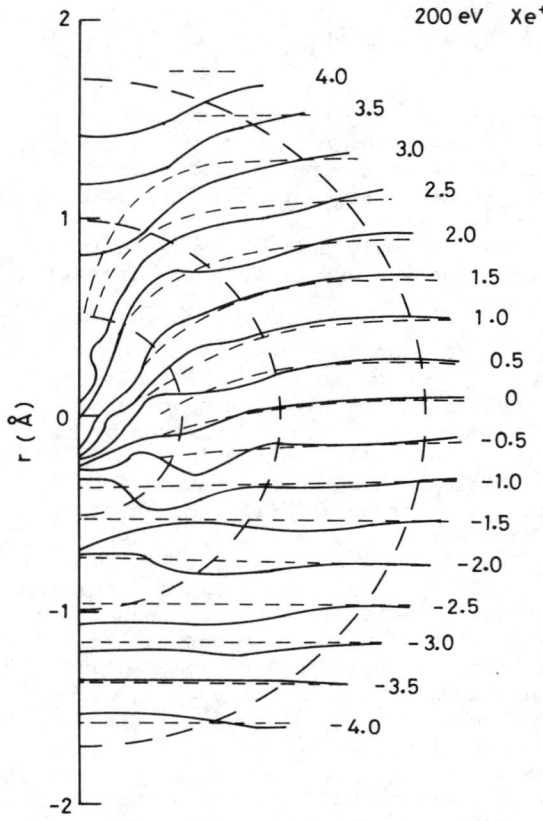

Fig. 9. Surfaces of equal phase (wave fronts) for the optical model wave functions (solid lines) for 200 eV electron scattering on xenon ions.

Fig. 10. The 400 eV coplanar symmetric (e,2e) experiment or Ugbabe and Weigold[17] for the 3p orbital or argon compared with the distorted-wave (full line) and plane-wave (dashed line) off-shell impulse approximations.

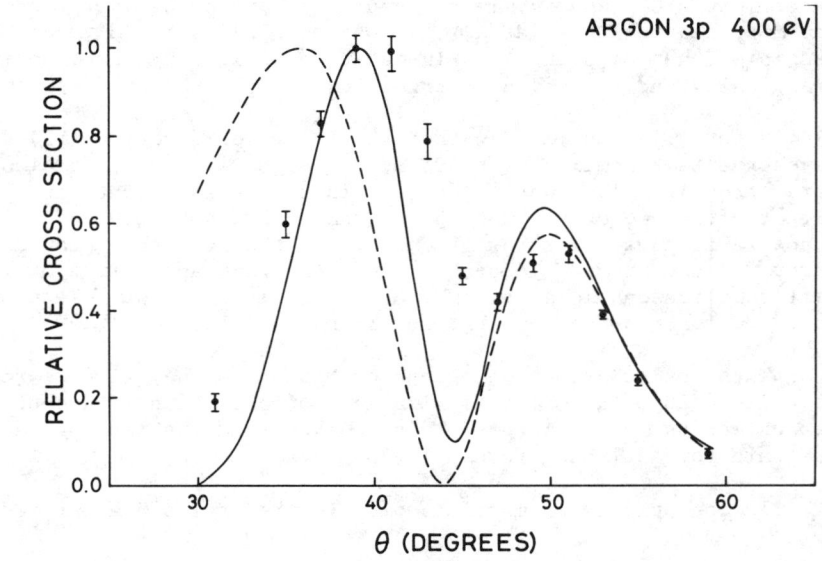

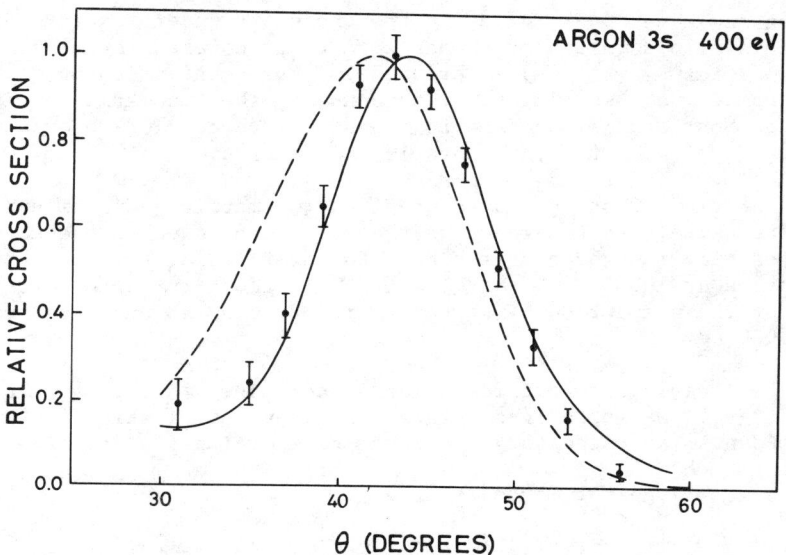

Fig. 11. The same as Fig. 9 for the 3s orbital of argon at 400 eV.

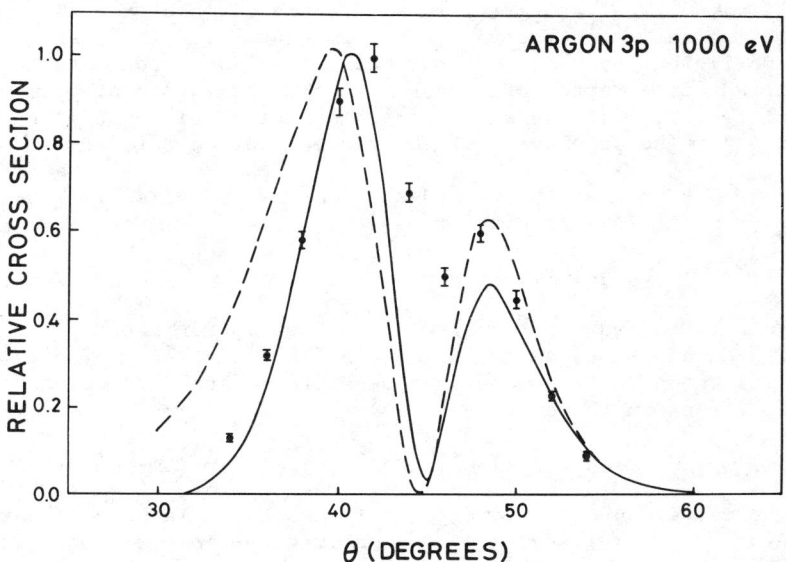

Fig. 12. The same as Fig. 9 for the 3p orbital of argon at 1000 eV.

an angle near 45° (42.3° in the present Flinders experiments). This allows values of q near zero to be covered for separation energies near 20 eV). The angle of one counter is varied. This geometry allows minimum disturbance to the kinematic conditions. The Mott scattering T-matrix changes imperceptibly over a wide range of ϕ, but q can be varied as desired.

Fig. 13 shows that noncoplanar symmetric q distributions are insensitive to the reaction theory. Plane waves are adequate to describe the shape of the momentum profile for the 3p orbital of argon at 1000 eV for q < 1 a.u. This geometry is therefore ideal for the extraction of structure information at energies near 1000 eV.

The structure information is contained in the overlap function $(f|g)$. We write these states as linear combinations of determinants $|\alpha\rangle$ using target Hartree-Fock orbitals ψ_j for the basis.

$$|g\rangle = \Sigma_\alpha a_\alpha^{(g)} |\alpha\rangle, \qquad (13)$$

$$|f\rangle = \Sigma_{j\beta} t_{j\beta}^{(f)} C_{jr\beta} \psi_j^\dagger |\beta\rangle \qquad (14)$$

$$(f|g) = n_r^{1/2} \Sigma_{j\alpha} a_\alpha^{(g)} t_{j\alpha}^{(f)} C_{jr\alpha} \psi_j 1 \qquad (15)$$

The final state $|f\rangle$ is written in terms of a basis consisting of a hole $\psi_j^\dagger$ in the target determinant $|\beta\rangle$. The Clebsch-Gordan coefficient ensures that the basis function belongs to the same irreducible representation r of the molecular point group as the state $|f\rangle$. The configuration-interaction (CI) coefficient is $t_{j\alpha}^{(f)}$. The degeneracy of the representation r is n_r.

For most of the closed-shell atoms and molecules so far studied it is sufficient to assume that

$$|g\rangle \cong |0\rangle. \qquad (16)$$

This is called the target Hartree-Fock approximation. In this case the Clebsch-Gordan coefficient in (15) is $n_r^{-1/2}$. The (e,2e) cross section is summed over degenerate final states, giving a factor n_r in the cross section:

$$\sigma = K n_r S_j^{(f)} |\langle \underline{k}'|T(p^2|\underline{k}\rangle|^2 |\langle \chi_A^{(-)} \chi_B^{(-)} | \psi_j \chi_0^{(+)} \rangle|^2, \qquad (17)$$

where K is an energy-independent constant. In noncoplanar symmetric geometry, $|\langle \underline{k}'|T(p^2)|\underline{k}\rangle|^2$ varies imperceptibly with ϕ. The quantity

$$S_j^{(f)} = [t_{j0}^{(f)}]^2 \qquad (18)$$

is called the spectroscopic factor. It is the square of the CI

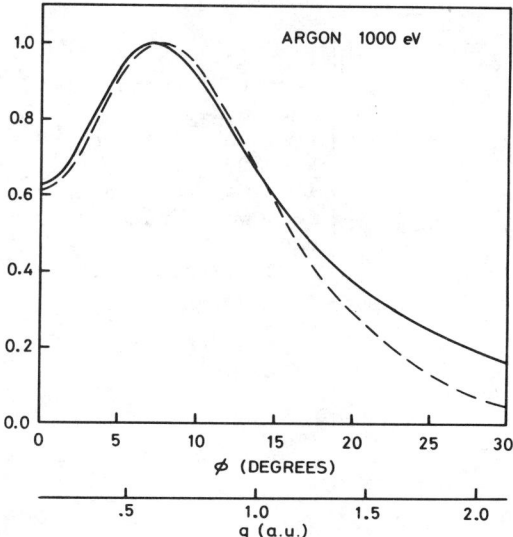

Fig. 13. The 3p orbital of argon at 1000 eV in noncoplanar symmetric geometry. The dotted curve is computed with plane waves. The full curve uses distorted waves.

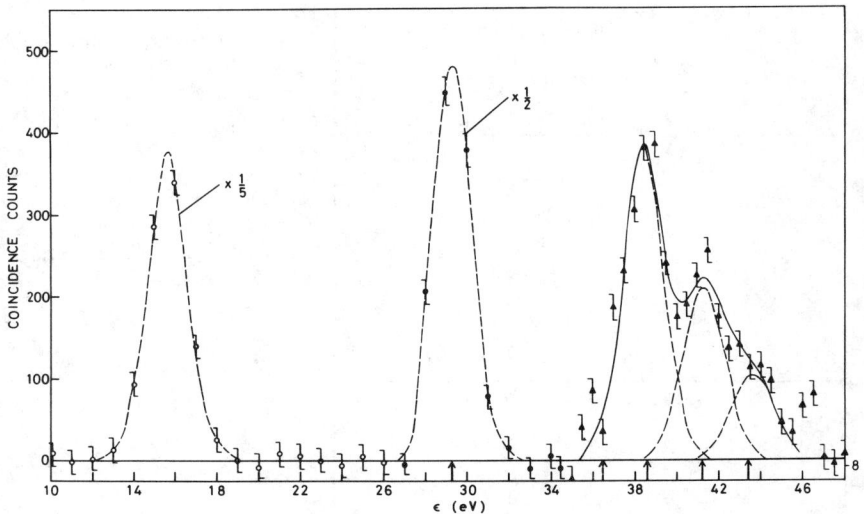

Fig. 14. The differential cross section for the 400 eV (e,2e) reaction on argon at $\phi = 0°$ plotted against separation energy ε.

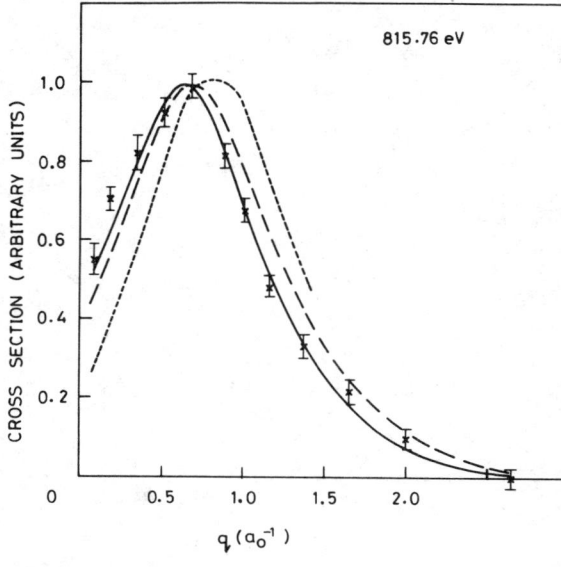

Fig. 15. The differential cross section for (e,2e) on the 15.76 eV level of argon at E = 800 eV. The full line uses the Hartree-Fock wave function, the dashed line uses Hartree-Fock-Slater and the dotted line uses the effective Coulomb function, $Z' = 7.517$.

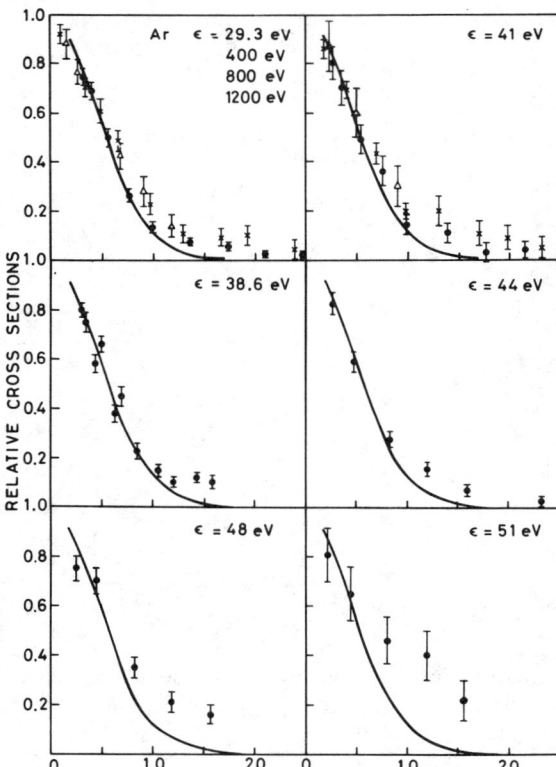

Fig. 16. Recoil momentum distribution for different 1/2 + ion states and continuum ($\varepsilon > 43.6$ eV) ion states in the noncoplanar symmetric (e,2e) reaction on argon at 1200 eV (filled circles), 800 eV (crosses) and 400 eV (triangles). The curve in every case is the 3s theory.

coefficient for the one-hole configuration in the sum (14), and as such obeys the sum rule

$$\Sigma_f S_j^{(f)} = 1. \tag{19}$$

This sum rule says essentially that the sum of all cross sections for states of the same symmetry as f is proportional to the number of electrons occupying the single-particle orbital ψ_j of that symmetry. It enables us to check our interpretation of the structure of a system.

The orbital energy is given by

$$E_j = <0|H|0> = \Sigma_f S_j^{(f)} \varepsilon_f. \tag{20}$$

It is the centroid of the eigenvalues for all states of the same symmetry as f.

We will illustrate the analysis of a system by (e,2e) spectroscopy using the example of argon. Fig. 14 shows the spectrum of separation energies ε for a fixed ϕ. The symmetry of each final state is identified by the q-distribution for each state. Cross sections for all states of the same symmetry are summed. The q-distribution for the 15.76 eV state as having 3s symmetry. The sum rule (19) in fact checks within the 10% experimental error confirming the analysis. The 3s orbital energy is determined to be 34.0±1.0 eV, compared to the HF value of 34.8 eV.

Spectroscopic factors for the argon eigenstates are as follows. They constitute vital information for testing a theory of electron correlation in argon, such as a CI calculation. The values are independent of incident energy, and therefore constitute hard structure information, which is not confused by possible reaction-mechanism dependence that is not properly understood.

Eigenvalue (eV)	Symmetry	$S_j^{(f)}$
15.76 ± .1	3p	1
29.3 ± .1	3s	0.58 ± .06
38.6 ± .1	3s	0.23 ± .02
41.2 ± .2	3s	0.13 ± .02
43.4 ± .1	3s	0.06 ± .02

For molecules there is no optical model available. We use plane waves at 1200 eV, since we have shown their validity for quite large atoms at this energy in the noncoplanar geometry.

Fig. 17 shows the analysis of ethane using the orbitals of Snyder and Basch[19], which are computed in an s,p gaussian basis. Shapes and magnitudes of q profiles are in excellent agreement with

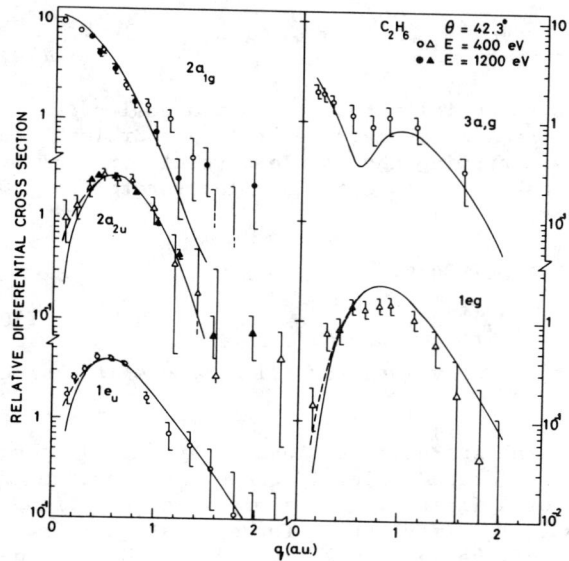

Fig. 17. The q-distribution for noncoplanar symmetric (e,2e) on ethane at 1200 eV.

theory.

Details of the identification of states and the understanding of CI coefficients in atoms and molecules will be given in a subsequent talk by Erich Weigold.

REFERENCES

1. O. Chamberlain and E. Segrè, Phys. Rev. 87, 81 (1952).
2. J.M. Wilcox and B.J. Moyer, Phys. Rev. 99, 875 (1955).
3. H. Tyrén, P. Hillman and Th.A.J. Maris, Nucl. Phys. 7, 10 (1958).
4. P. Hillman, H. Tyrén and Th.A.J. Maris, Phys. Rev. Letters 5, 107 (1960).
5. G. Jacob and Th.A.J. Maris, Rev. Mod. Phys. 38, 121 (1966).
6. J.P. Garron, J.C. Jacmart, M. Riou, C. Ruhla, J. Teillac and K. Strauch, Nucl. Phys. 37, 126 (1962).
7. Y. Sakamoto, Prog. Th. Phys. 28, 803 (1962).
8. K.L. Lim and I.E. McCarthy, Phys. Rev. 133, B1006 (1964).
9. N.D. Birrell, I.E. McCarthy and C.J. Noble, unpublished.
10. S.K. Young and E.F. Redish, Phys. Rev. C 10, 498 (1974).
11. H.V. v Geramb and D. Eppel, Z. Phys. 261, 177 (1973).
12. G.A. Baker, Jr., I.E. McCarthy and C.E. Porter, Phys. Rev. 120, 254 (1960); Yu.F. Smirnov and V.G. Neudatchin, JETP Lett. 3, 192 (1966); A.E. Glassgold and G. Ialongo, Phys. Rev. 175, 151 (1968).
13. U. Amaldi, Jr., A. Egidi, R. Marconero and G. Pizzella, Rev. Sci. Instr. 40, 1001 (1969).
14. R. Camilloni, A. Giardini-Guidoni, R. Tiribelli and G. Stefani, Phys. Rev. Lett. 29, 618 (1972).
15. E. Weigold, S.T. Hood and P.J.O. Teubner, Phys. Rev. Letters 30, 475 (1973).
16. J.B. Furness and I.E. McCarthy, J. Phys. B.6, 2280 (1973) and I.E. McCarthy, C.J. Noble, B.A. Phillips and A.D. Turnbull, unpublished.
17. A. Ugbabe and E. Weigold, unpublished.
18. C.J. Noble, unpublished.
19. L.C. Snyder and H. Basch, <u>Molecular Wave Functions and Properties</u>, Wiley (N.Y.) 1972.

INTERMEDIATE ENERGY NUCLEAR REACTIONS

G. E. Walker[*]
Indiana University, Bloomington, In. 47401

ABSTRACT

Intermediate energy electron, photon, pion, kaon, and proton reactions are discussed and compared. Special emphasis is placed on the complementarity of the various probes. Production of an intermediate nucleon resonance is proposed as a mechanism for understanding the recent medium energy (γ,p) results.

INTRODUCTION

The main objectives in this talk are:
1) To review the nuclear situation in a general way with respect to the kinds of things one hopes to learn from simple medium energy reactions, the difficulties, and possible schemes for overcoming the problems. My hope is to use this approach to inform researchers in atomic and molecular knockout reactions and at the same time set the stage for more specialized talks by the nuclear theorists and experimentalists.
2) To present some new results or ideas involving
    a) knockout reactions involving two step processes and intermediate Δ's, and
    b) a comparison of different probes including kaons, protons, pions, and electrons as specific examples of some of the more recent developments in medium energy reactions.

The nucleus is a finite many-body system held together by strong forces. We believe the nucleus is primarily composed of "dressed" neutrons and protons interacting via a strong force which bears a reasonable resemblance to that studied in a narrow kinematic region in free nucleon-nucleon scattering. This interaction is characterized by a short range potential that is attractive with a repulsive core. This potential results in a single particle momentum distribution that is conjectured to have a high momentum tail. For ease of visualization and calculation, we most often picture the nucleons as occupying well defined orbits in the nucleus consistent with the Pauli exclusion principle. An educated "first guess" for the single particle orbitals is obtained by doing a Hartree-Fock calculation using a nucleon-nucleon interaction similar to that discussed above. In studying intermediate energy (p,2p) reactions, one has the possibility of obtaining information relevant for the picture summarized above. As examples one should learn about

---

[*]Research supported in part by the National Science Foundation.

NUCLEAR STRUCTURE

1) the binding energy, momentum distribution, occupation probability, and width of the shell model orbitals,
2) the evidence for nuclear constituents other than nucleons, such as Δ's and N's.

REACTION MECHANISM

3) the medium-energy nucleon-nucleus reaction mechanism (this can be helpful in unraveling other processes).

Of course the ability to obtain structure information is severely hampered by
1) uncertainties in the form of the off-shell effective t matrix,
2) multistep processes, "final state" interactions (distortion), three body effects,
3) effects of anti-symmetrization, correlations, and Δ production.

To overcome some of these difficulties, one should do medium energy - high resolution experiments with a particular probe varying the energy, state reached, reaction, and target. One can also use different probes. Since each probe has its own particular characteristics and limitations, the comparison of results obtained with different probes can be particularly fruitful.

COMPARISON OF DIFFERENT PROBES

First, let us consider in more detail some of the probes expected to be available from the new generation of medium energy facilities. We concentrate on the electron, proton, pion, and kaon probes. We shall review the basic properties of these probes and compare the different kinds of results they are expected to give for inelastic scattering.

ELECTRON SCATTERING

The basic probe-nucleon interaction is well understood, except at high momentum transfer where the effects of meson-exchange currents is a messy problem. Because of the presence of a $\tan^2 \theta/2$ factor, the transverse form factor is the important contributor to electron scattering at large angles.[1] This means, for medium energy electrons at high momentum transfer, one is studying the convection current and magnetization density of the nucleus. Because the isovector magnetic moment dominates the transverse form factor, one preferentially excites $\Delta T = \Delta S = 1$ states at medium and high momentum transfer, assuming a $T = 0$ major closed shell ground state. Some selected experimental results and comparison with theory are shown in Figs. 1[2] and 2.[4] In general, the agreement between theory and experiment is good for the high spin $T = 1$ states strongly excited at large momentum transfer.[2,3,4]

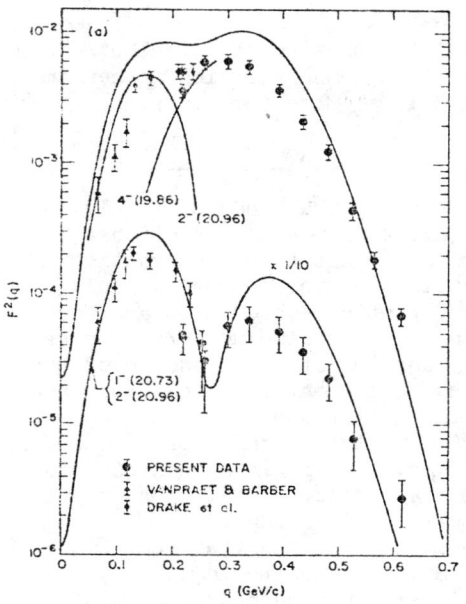

Fig. 1. Form factors $F^2(q) = \dfrac{d\sigma/d\Omega}{[4\pi\sigma_M(\frac{1}{2}+\tan^2\frac{\theta}{2})]}$ for selected t=1 states (energies in parenthesis) in ^{16}O excited in inelastic electron scattering at 135°. The theoretical predictions and data denoted "present data" shown are taken from reference 2. The data denoted "Vanpraet and Barber" ("Drake et al.") is from reference 11 (12).

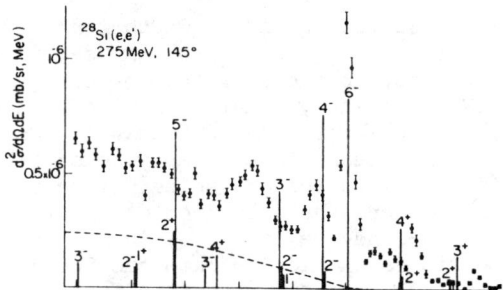

Fig. 2. The cross section $d^2\sigma/d\Omega dE$ for inelastic electron scattering from ^{28}Si at E_e=275 MeV, θ=145° unfolded for radiative process. The non-negligible cross sections predicted by the particle-hole model for q=525 MeV/c are shown as spikes (arbitrary overall scale). The dashed line is the computed quasi-elastic spectrum. The data and theoretical predictions are taken from reference 4.

We shall see later that, even for electromagnetic knockout reactions, such as (γ,p), the reaction mechanism is uncertain at medium energies and large momentum transfer.

HADRONIC SCATTERING

Now before comparing the kinds of results obtained or expected from proton, pion and kaon-nucleus reactions, it's useful to review the basic probe-target constituent interaction. In particular note
 1) antisymmetrize in nucleon-nucleus only
 2) for mesons, particle absorption and reemission and crossing symmetry may be important to include from the beginning or one may confuse the energy and momentum dependence of the basic interaction.

If we examine the elementary total cross sections for these hadronic probes, the difference in the energy dependence is striking. For example the pion-nucleon total cross section is rapidly increasing from 50 to 200 MeV, the nucleon-nucleon total cross section is decreasing, and the K^+ nucleon total cross section is nearly constant and relatively small. This allows
 1) nuclear mean free path differences
 2) different multiple scattering theory validity range
 3) different energy dependence of multistep processes for the various probes.

KAON (K^+) REACTIONS

Note that the K^+ (kaon) (unlike the K^-) is an ideal probe for studying nuclear structure. The basic K^+ + nucleon interaction is weak, slowly varying with energy, spin independent, and S wave for $p_{K^+}^{lab} \leq 350$ MeV/c. This means multiple scattering theory and the usual approximations adopted should be demonstrably valid for a change. Although there are proposals to upgrade existing facilities, my understanding is that, in the near future, energy resolution of better than one MeV is unlikely. But even with this limitation, I feel that K^+ induced nuclear reactions (including knockout of nucleons) occupying valence and more deeply bound orbitals represent a useful and exciting new source of nuclear structure information that will nicely complement the data from other probes. Because of the spin and isospin dependence of the basic K^+ nucleon interaction normal parity ΔT=0 states should dominate the inelastic scattering response function at all momentum transfers for $p_{K^+}^{lab} \leq 350$ MeV.[5]

PROTON REACTIONS

The proton-nucleon reaction (which will be discussed in considerable detail at this conference) is quite complicated. The effective interactions adopted to study inelastic scattering to date have the property that the short range part is both spin and isospin dependent, so that for high momentum transfer inelastic proton scattering from a T=0 nucleus, non-normal parity T=1 states are predicted to dominate the spectrum. In Fig. 3, we show some recent (p,p') data on ^{28}Si from IUCF.[6] A 6^-, T=1 state predicted to dominate the spectrum at high momentum transfer is identified with a prominent experiment peak observed. T=0 excited states are present at large q because of exchange effects also predicted by theory.

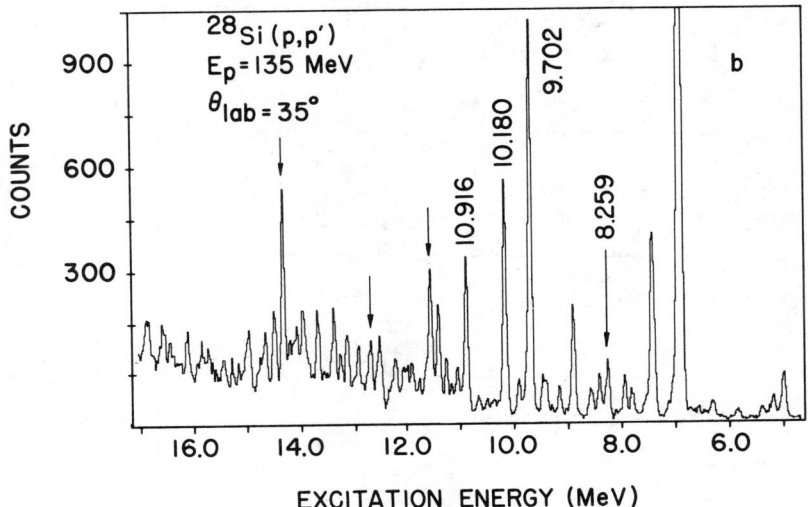

Fig. 3. Spectrum for the reaction ^{28}Si(p,p')^{28}Si taken with 135-MeV protons. Some of the peaks used for energy calibration are identified by their excitation energies. The state denoted by arrow at ~14 MeV is the 6^- T=1 state predicted to be prominent by theory and seen also in inelastic electron scattering. The data is taken from reference 6.

PION REACTIONS

The study of (π,π') and $(\pi,\pi'n)$ both theoretically and experimentally has just gotten under way in the last couple of years. It is not yet clear whether a DWIA approach with a separable form for off-shell t matrix momentum space form factor will be adequate for studying pion reactions. It may be that a field theoretic approach including crossing and explicit boson emission and absorption is required. Our understanding of the pion nucleus reaction mechanism is significantly less developed than the other probes I have discussed this morning. For the purposes of comparison with other probes, we now present results for pion inelastic scattering at 70 and 180 MeV from ^{16}O.[7]

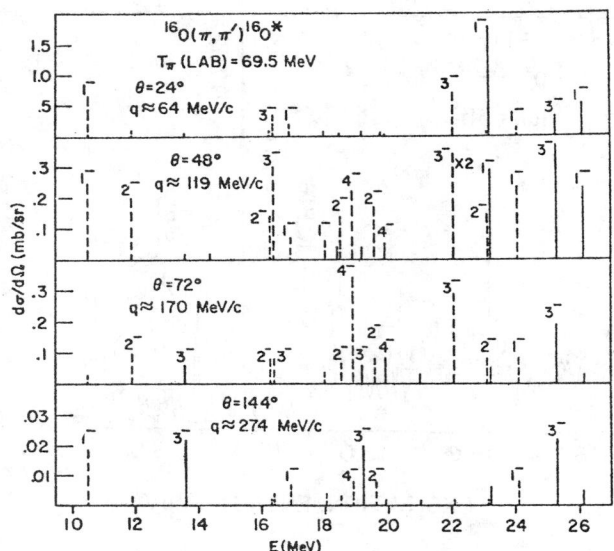

Fig. 4. Calculated pion-^{16}O inelastic scattering differential cross sections as a function of the final nuclear excitation energy E. The initial lab kinetic energy of the pion T_π(lab) is 69.5 MeV. The differential cross sections are shown for four different scattering angles (momentum transfers). Solid lines correspond to T=1 final nuclear excited states (obtained using the particle-hole model in the Tamm-Dancoff approximation) while T=0 states are represented by dotted lines. The spin and parity J^π of the more prominently excited states is indicated. Only states with appreciable cross sections are included. The predictions are taken from reference 7.

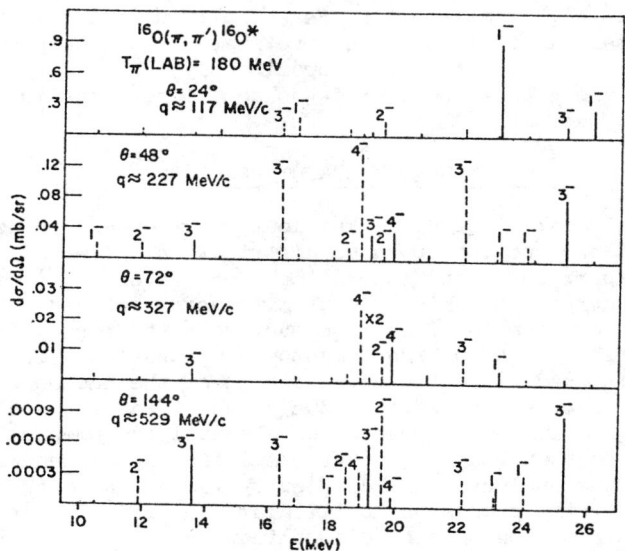

Fig. 5. Pion inelastic scattering cross sections as in fig. 4 except $T_\pi(lab)$ is 180 MeV. The predictions are taken from reference 7.

Because of the basic properties of the pion-nucleon interaction, T=0 high spin non-normal parity states dominate the spectrum at large momentum transfer (note T=1 states dominated for the protons and electrons). As mentioned earlier, if one carries out a similar calculation for K^+ inelastic scattering, T=0 high spin normal parity states dominate the spectrum at high momentum transfer. In addition to the fact that the various probes characteristically excite different kinds of states at high momentum transfer, it is useful to note that the low spin normal parity T=0 "collective states" and the giant dipole states are predicted to appear prominently in the nuclear response for all the probes considered--so some important direct comparisons will be possible.

PHOTON INDUCED KNOCKOUT REACTIONS

One important goal of studying knockout reactions is to obtain information about the nucleon momentum distribution. I would like to discuss recent experimental and theoretical results on (γ,p) reactions at medium energies and high momentum transfer because this reaction provides a nice example of how one must come to grips with and might separate effects due to high momentum components and multistep processes both with and without excited states of the nucleon.

Clearly, in order to see effects due to short range correlations (which induce high momentum components in the nucleon momentum distribution), one looks at reactions that result in high momentum transfer delivered to the nucleus. The difficulty is that at high momentum transfer, multistep processes (particularly hadronic) must be investigated as a reaction mechanism complication. Note that in considering a particular multistep process, some care must be taken not to have, in principle, double counted, because the diagram may already be included in obtaining the optical potential for distorted waves. If one works at fixed momentum transfer and varies the energy of the incoming probe, qualitatively, the importance of two step processes compared to a one step mechanism can be obtained by looking at the energy dependence of the probe-nucleon interaction.

First we consider the (γ,p) data shown in Fig. 6.[8] The higher energy points are from Matthews et al. at M.I.T.[8] The fact that, as the energy and momentum transfer increases the cross section appears to increase, implied to Londergan, Nixon and myself[9] that the usual multistep processes or hardcore induced high momentum components are not the most important feature of the effect. Our suggestion is that intermediate Δ production is the explanation. If it is, then it will have to be reckoned with in medium energy knockout processes.

The basic diagram we consider in addition to the one step process is shown in Fig. 7.[9]

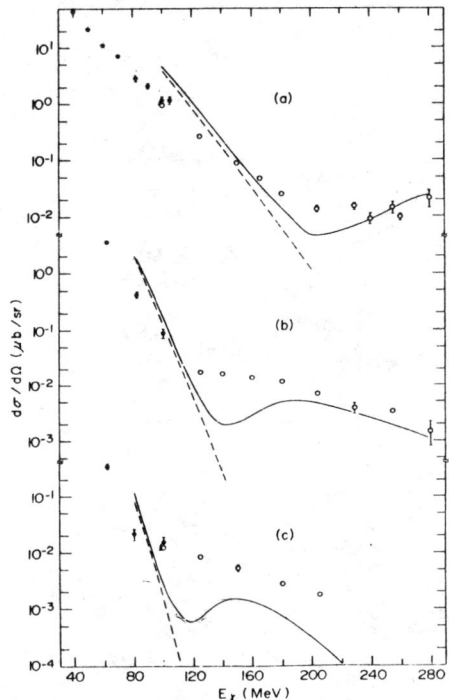

Fig. 6. Differential cross section in the laboratory system for the reaction $^{16}O(\gamma, p_o)^{15}N$ as a function of photon energy at proton angles (a) 45°, (b) 90°, and (c) 135°. Solid circles are used for the data of reference 10, open circles for the results presented in reference 8. Only statistical errors are shown. The curves are theoretical calculations taken from reference 9. The dashed curve, single-step process only; solid curve one step plus two step process (see Fig. 7).

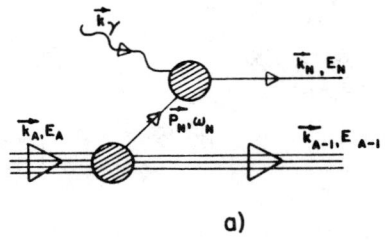

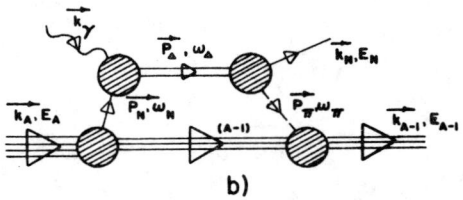

Fig. 7. Diagrammatic representation of two mechanisms for photo-proton emission with initial photon momentum $\vec{k}_\gamma$ and nuclear momentum $\vec{k}_A$ leading to final proton momentum $\vec{k}_N$. (a) Direct, or one-step process. The proton is knocked out by interaction with the photon. (b) Two-step contribution due to isobar formation. A nucleon (proton or neutron) is converted by the photon into a $\Delta(1232)$, which propagates and then decays into a proton plus pion, the pion being absorbed by the residual nucleus.

The results of the calculations are shown in figure 6. The agreement with experiment is encouraging; however, effects of distortion and short range correlations have not yet been included. A rather extensive program to include these as well as studying the effects of ρ exchange and giving the Δ a width is currently planned by Londergan and Nixon.

In conclusion, there are many possible complications that must be dealt with in studying knockout reactions. If the desirec information is only obtainable by including high momentum transfer knockout reactions, one may have to consider multistep processes involving Δ production. As in the earlier discussion involving different probes and inelastic scattering, the use of γ, e, π, n, and K^+ probes together is likely to be necessary in order to have a chance to unravel the structure and reaction mechanism uncertainties.

REFERENCES

1. T. DeForest, Jr. and J. D. Walecka, Advan. Phys. 15, 1 (1966).
2. I. Sick, E. B. Hughes, T. W. Donnelly, J. D. Walecka, and G. E. Walker, Phys. Rev. Lett. 23, 1117 (1969).
3. T. W. Donnelly and G. E. Walker, Ann. Phys. (N.Y.) 60, 209 (1970).
4. T. W. Donnelly, J. D. Walecka, G. E. Walker, and I. Sick, Phys. Lett. 32B, 545 (1970).
5. G. E. Walker, B.A.P.S. 21, 646 (1976).
6. G. S. Adams, A. D. Bacher, G. T. Emery, W. P. Jones, R. T. Kouzes, D. W. Miller, A. Picklesimer, and G. E. Walker, submitted to Phys. Rev. Lett.
7. M. K. Gupta and G. E. Walker, Nucl. Phys. A256, 444 (1976).
8. J. L. Matthews, W. Bertozzi, M. J. Leitch, C. A. Peridier, B. L. Roberts, C. P. Sargent, W. Turchenetz, D. J. S. Findlay, and R. O. Owens, Phys. Rev. Lett. 38, 8 (1977).
9. J. T. Londergan, G. D. Nixon, and G. E. Walker, Phys. Lett. 65B, 427 (1976).
10. R. O. Owens, private communication; D. J. S. Findlay, thesis, Glasgow University (1973) and to be published.
11. G. J. Vanpraet and W. C. Barber, Nucl. Phys. 79, 550 (1966).
12. T. E. Drake, E. L. Tomusiak, and H. S. Caplan, Nucl. Phys. A118, 138 (1968).

(p,2p) EXPERIMENTS AT THE
UNIVERSITY OF MARYLAND CYCLOTRON*

P. G. Roos
University of Maryland, College Park, Md. 20742

ABSTRACT

This paper discusses recent (p,2p) experiments at the Maryland Cyclotron, and the theoretical analyses. In addition DWIA results are presented which help to establish guidelines for future (p,2p) studies.

INTRODUCTION

I will be discussing some of the (p,2p) work which has been carried out by myself and my colleagues at the Maryland Cyclotron. I will begin by briefly discussing the reaction, the types of experimental techniques and the various theoretical treatments presently available to analyze (p,2p) reaction data. Secondly, I will present experimental and theoretical studies of (p,2p) on 1 S-shell and 1 P-shell nuclei carried out by the Maryland group. Finally, I want to specifically discuss the effects of distortion on the experimental data by presenting theoretical calculations for ^{12}C, ^{40}Ca and ^{208}Pb at various bombarding energies. In the (p,2p) reaction a measurement of the two outgoing protons (energies and angles) provides a complete determination of the three body final state, i.e., the momentum $\vec{p}_3$ of the core. Even if the core is unbound, leading to a four-body final state, the measurement of the two protons still determines the center-of-mass momentum of the core particles and their relative energy.

Experimentally, one places detectors at a pair of angles and measures the energies (E_1, E_2) of the detected coincident protons. In a contour plot of E_1 vs. E_2 all events for a specific final state in the core lie on a particular 3-body kinematic locus. The distribution of events along these loci reflects the reaction mechanism and therefore the "physics." For each point on the $E_1 - E_2$ plane the recoil momentum and therefore recoil energy E_3 is determined. Thus, by calculating E_3, we can use the data to form the binding energy spectrum (BE=$E_0-E_1-E_2-E_3$). This presentation of the data clearly shows the relative excitation of the various final states in the core. For any particular final state the data can be projected onto the E_1 proton energy axis forming the triple differential cross section, which is generally referred to as an energy sharing spectrum. These data may also be plotted as a function of recoil momentum instead of E_1, since the recoil momentum is uniquely determined.

*A lengthier discussion can be found in the University of Md. Tech. Rep. No. 77-038.

Now let me turn to the theory - at least the theory of direct proton knockout which we hope is applicable in one form or another. Fig. 1 shows the first order knockout diagram, which although it is a tremendous oversimplification, we still hope that this diagram dominates the reaction mechanism. Considerations based on this diagram do provide and have provided most of the experimental ideas for testing the reaction mechanism. The lower vertex represents the virtual break-up of the target nucleus into a proton and the core in some particular state, and thus contains the target nuclear structure information. The upper vertex represents an off-the-energy shell interaction of two protons. If this diagram were applicable, one can clearly see the great wealth of nuclear physics available in such (p,2p) studies, i.e., both nuclear structure information about target nuclei, and nucleon-nucleon interaction information.

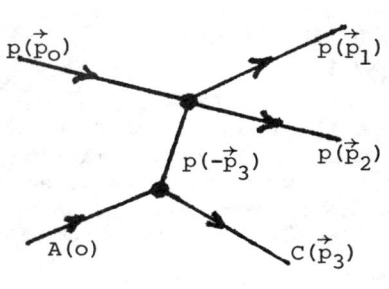

Fig. 1. First order diagram

In the plane wave impulse approximation (PWIA) or spectator model the cross section for this diagram can be written as:

$$\sigma(\theta_1,\theta_2,E_1) = (PSF) \times \sigma_{pp}(\theta) \times |\phi(-\vec{p}_3)|^2$$

where (1)

$$\sigma_{pp}(\theta) \alpha |<\vec{p}_1\vec{p}_2|t|\vec{p}_0,-\vec{p}_3>|^2 \text{ and } \phi(-\vec{p}_3) \alpha \int e^{i\vec{p}_3 \cdot \vec{R}} <\psi_C|\psi_A> d^3R$$

PSF is a kinematic factor consisting of known quantities. The cross section $\sigma_{pp}(\theta)$ in this factorized form is properly a half-shell cross section. Generally one replaces $\sigma_{pp}(\theta)$ by some on-shell prescription, the two most popular being the initial energy prescription ($\sigma_{pp}(\theta)$ is taken to correspond to the initial state of the two protons before the collision), and the final energy prescription ($\sigma_{pp}(\theta)$ is taken to correspond to the relative energy and angle of the two protons in the final state). However, one can calculate the half-shell cross section for various nucleon-nucleon potentials. Redish, Stephenson, and Lerner[1] have shown that these prescriptions and the half-shell cross section can differ significantly - particularly at lower energies. I will show some of these effects later when comparing to experimental data.

The quantity $\phi(-\vec{p}_3)$ represents the Fourier transform of the overlap integral between the initial and final nuclei, and is therefore simply the single particle wave function in momentum space. The simplicity of this expression arises from the assumption of non-interaction with the core. It would be nice if this diagram were adequate; however, it has been shown many times that one cannot ignore the interaction of the protons with the core, even for d(p,2p)n. Thus, we need a more sophisticated theory. One possible extention is to a multiple scattering theory[2] where one explicitly includes double scattering, triple scattering, etc. This procedure has been carried out for deuterium using the Fadeev equations as I will show later. However, at this time the procedure is too difficult to carry out any very realistic calculation.

The most direct, and numerically simplest procedure, is to add distorted waves in order to take account of the interaction of the protons with the core. In the distorted wave impulse approximation (DWIA)[3,4,5] one retains the factorization approximation. Thus the three-body cross section expression maintains the same form as the PWIA (Eq. 1). However, ϕ is now the so-called distorted momentum distribution

$$\phi \alpha \int \chi_1^{(-)*}(\vec{r}) \chi_2^{(-)*}(\vec{r}) U(\vec{r}) \chi_o^{(+)}(\gamma\vec{r}) d^3r$$

where the χ's are distorted waves for the incoming and outgoing protons calculated using potentials which fit elastic scattering and $U(\vec{r})$ is the bound proton single particle wave function. A number of DWIA calculations have been carried out over the years and provide rather good fits to the experimental data. These calculations clearly show the necessity of including the distortion introduced by the core.

Finally, if a suitable representation of the nucleon-nucleon interaction in coordinate space could be found, one can remove the factorization approximation, put the interaction into the calculation, and carry out the full 6-dimensional calculation, a rather time consuming computation compared to DWIA.

$$\sigma(\theta_1,\theta_2,E_1) \alpha \left| \int\int \chi_1^{(-)*}(\vec{r}_1) \chi_2^{(-)*}(\vec{r}_2) t(\vec{r}_1,\vec{r}_2) U(\vec{r}_2) \chi_o^{(+)}(\vec{r}_1) d^3r_1 d^3r_2 \right|^2$$

Such calculations (DWTA) have been carried out by McCarthy[6] using a pseudo-potential which reproduced the energy dependence of the 90° p-p scattering cross section. Unfortunately this pseudo-potential does not reproduce the angular dependence of the p-p cross section, and therefore one questions its general applicability. Later I will present some comparisons of the DWIA and DWTA to show the accuracy of the factorization approximation - at least within the framework of this particular pseudo-potential. However in my opinion one should abandon this coordinate space pseudo-potential for (p,2p) and work in momentum space. Calculations such as those of Koshel[7] where the distorted waves are expanded in a plane wave set should be ideal for properly investigating the factorization approximation.

From an experimental standpoint the great advantage of 3-body breakup reactions such as (p,2p) compared to the equivalent transfer reaction lies in their greater flexibility. For example,

independent of the bombarding energy one can always choose angles
to provide momentum matching, so that one can investigate the target nucleon with zero momentum. Thus it is possible to choose a
high energy to minimize distortion, and still investigate target
nucleon momenta from 0 MeV/C to some very large value. In addition,
the flexibility of a second angle allows one to make a much more
detailed investigation of the reaction mechanism.

Fig. 2 shows three typical experimental studies of the (p,2p)
reaction (or any 3-body reaction). These studies are based on
the dominance of the first order diagram (or the DWIA equivalent).
The first experiment is the energy sharing experiment in which the
detectors are fixed at equal proton angles such that zero recoil momentum is kinematically allowed (quasi-free angles). One measures
the cross section $\sigma(\theta,-\theta,E_1)$ as a function of the energy of one
proton. For this particular arrangement the two-body cross section
does not vary dramatically in energy or angle over the range of
energy sharing, and thus the structure of the cross section is
dominated by ϕ^2 (the distorted momentum distribution). I present
what one might typically expect for $L = 0$ and $L \neq 0$ proton knockout. Thus, in this experiment one is primarily isolating the lower
vertex.

The second experiment which I call a quasi-free angular distribution, attempts to isolate the upper vertex. In this experiment one measures the 3-body cross section for a variety of quasi-
free angles. Assuming distortion effects are not too angular and
energy dependent (or at least that we can calculate them) so that
for a fixed value of recoil momentum p_3, ϕ^2 is approximately constant, one can plot $\sigma(\theta_1,\theta_2,E_1)$/PSF versus the 2-body p-p center-of-

Fig.2. Types of (p,2p) experiments based upon $\sigma(\theta_1,\theta_2,E_1)$=PSF $\times \sigma_{pp}(\theta) \times |\phi(-p_3)|^2$.

1) Energy Sharing
 a) $\theta_1 = \theta_2$ fixed
 b) $\vec{p}_3 = 0$ kinematically allowed

2) Quasi-free angular distribution
 a) θ_1 and θ_2 varied, but chosen to allow $p_3 = 0$.
 b) $\sigma(\theta_1,\theta_2,E_1)[p_3=0]$/PSF extracted and plotted vs. θ_{pp}

3) Coplanar symmetric angular distribution
 a) $\theta_1 = \theta_2$ varied
 b) $|\vec{p}_1| = |\vec{p}_2|$
 c) Extract $\sigma(\theta_1,\theta_2,E_1)$ for $|\vec{p}_1| = |\vec{p}_2|$

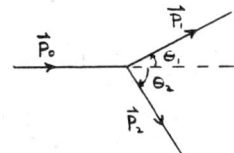

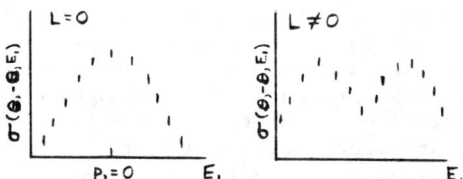

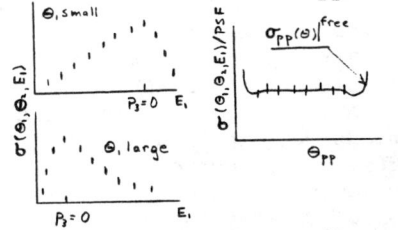

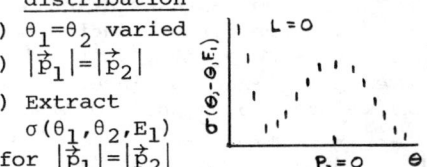

mass scattering angle. In the impulse approximation this quantity should be proportional to $\sigma_{pp}(\theta)$. In the sketch I have indicated what one expects for PWIA. This experiment can provide a rather nice test of the factorization approximation, especially since the flatness in angle of the p-p scattering cross section must arise from a delicate cancellation of the p and d waves, and has been rather successfully applied to α-particle knockout.[8,9,10] Unfortunately, to my knowledge, with the exception of d(p,2p) and d(p,pn) such a (p,2p) experiment has never been performed, and clearly should be.

Finally I show the canonical arrangement of the coplanar-symmetric angular distribution. In this experiment one measures the (p,2p) cross section for $\theta_1 = -\theta_2$ and $E_1 = E_2$ for a variety of equal angle pairs. Each angle corresponds to a different value of recoil momentum which is either parallel or anti-parallel to the incident beam direction. Again considering this study from the 1st order diagram we see that both vertices vary as a function of angle, the lower vertex because of the changing recoil momentum and the upper vertex because of the changing relative energy of the two protons. In this arrangement the 2-body p-p c.m. scattering angle is always 90°. However, the p-p relative energy can change dramatically. Schematically I have indicated what one might expect to see for the knockout of an s-state proton. At the larger angles the two-body cross section is fairly constant, since $\sigma_{pp}(90°)$ is rather flat at higher energies. Thus the structure is dominated by ϕ^2. However, at the smaller angles the relative energy between the two protons gets smaller therefore dramatically increasing the 2-body cross section, as I have indicated. In this small angle region one might hope to gain information on the off-shell behavior of the p-p cross section, since the reaction can be very far off-shell. However, many other effects such as distortion are extremely important in this region. Although this is the canonical experiment, I hope to convince you later that this is not the experiment to do first.

A = 2,3,4

Now let me turn to some experimental results for d, ^{3}He, and ^{4}He. First we consider d(p,2p)n and d(p,pn)p, which I treat separately since we do have multiple scattering calculations, as opposed to DWIA for the other cases. Fig. 3 shows the energy sharing cross sections for d(p,2p)n at the quasi-free equal angle pair for three energies. The curves are PWIA calculations using a Hulthen wave function normalized to the data by the factor indicated in the figure. Clearly the PWIA is inadequate even for d(p,2p)n being too large by a factor which is energy dependent, and not providing a terribly good representation of the shape. While at Maryland John Wallace performed exact 3-body multiple scattering calculations[11] for these data. His calculations used only a separable s-wave nucleon-nucleon interaction but the strength was modified to reproduce the on-shell nucleon-nucleon cross section. Using this interaction the full multiple scattering series was summed. Let me just show one example. Fig. 4 shows the (p,2p) and (p,pn) energy

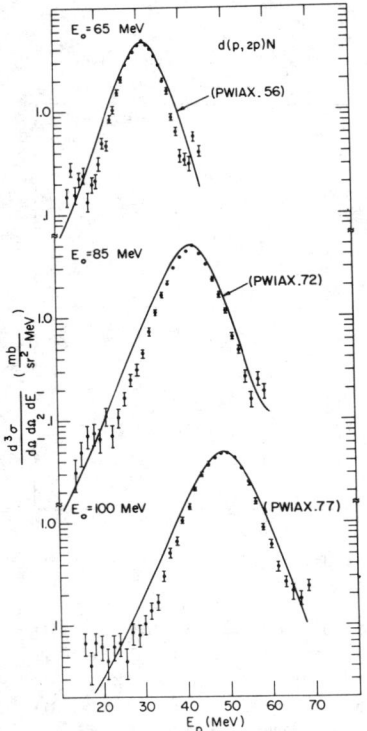

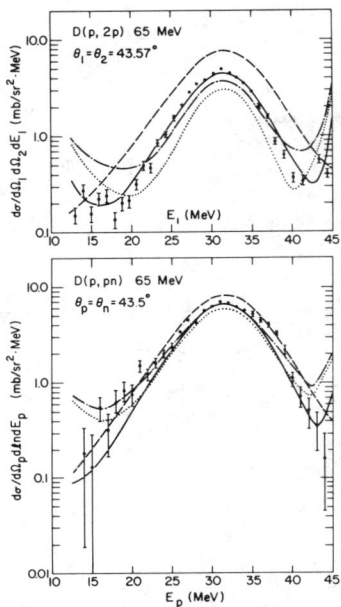

Fig. 4. Energy sharing spectra at 65 MeV. The curves are multiple scattering calculations (single, dashed; double, dash-dot; triple, dotted; full multiple scattering, solid).

Fig. 3. Energy sharing spectra for d(p,2p)n at equal symmetric angles. The curves are PWIA normalized to the data by the factor indicated.

sharing results at 65 MeV. Again one sees the inadequacy of single scattering--particularly for (p,2p). The inclusion of the full multiple scattering series provides extremely good agreement with the data. Similarly at 85 MeV and 100 MeV multiple scattering calculations provide excellent agreement with experiment.[12] In Fig. 5 I show some quasi-free angular distribution[12] results. Again one sees the tremendous improvement gained by including multiple scattering, but we also see significant discrepancies. One can at least qualitatively understand them as arising from (a) the neglect of the Coulomb interaction and (b) the use of pure s-wave for the p-n interaction (i.e., the p-n cross section is not flat as a function of angle as is the case with p-p scattering). Further results can be found in Ref. 12. In order to do any more we await a three-body calculation which includes higher nucleon-nucleon partial waves. In fact in my opinion experiment is so far ahead of three-body theory that we can wait several more years before providing any additional data on deuterium.

Now let me turn to ^{3}He, ^{4}He(p,2p). Fig. 6 shows the energy sharing cross sections for d, ^{3}He and ^{4}He at 100 MeV.[13] The curves are PWIA calculations normalized to the data by the factors indicated. One sees the pronounced increase in width as the binding

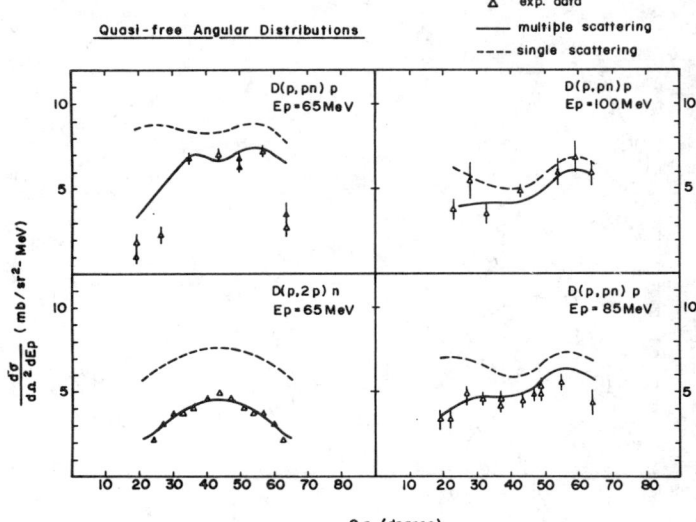

Fig. 5. Quasi-free angular distributions. θp is the 2-body scattering angle.

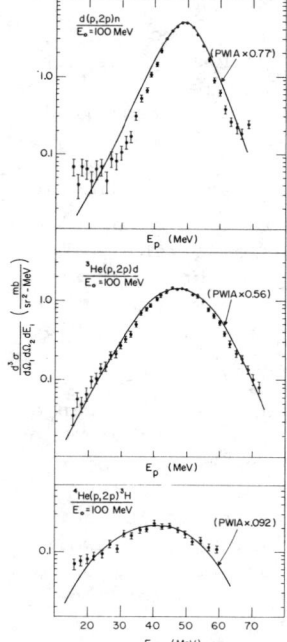

Fig. 6. Energy sharing spectra for (p,2p) on ^{2}H, ^{3}He, and ^{4}He. The curves are normalized PWIA calculations.

energy increase to 20 MeV for ^{4}He. Fig. 7 shows the ratio of the experimental cross section to the PWIA calculation as a function of energy. This clearly shows the reduction in multiple scattering effects as the energy increases, and the increase in multiple scattering as the A of the target increases, although detailed comparisons are not valid due to the inadequacy of the wave functions used - particularly for ^{4}He.

The data points were obtained using the final energy prescription for $\sigma_{pp}(\theta)$. If one uses the half-shell cross section instead, one obtains the curves represented by the dashed lines. Only for ^{4}He which is bound by 20 MeV does the 1/2-shell cross section make an important difference, the difference decreasing with increasing energy.

The rapid energy dependence of ^{4}He, plus a Phys. Rev. Letter[14] which reported a fit to the ^{4}He data with the extremely simplified attenuation model prompted me to try DWIA calculations for ^{4}He. Without going into details I performed DWIA calculations[15] as a function of energy using

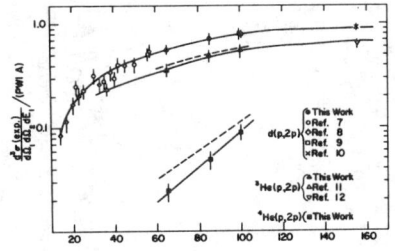
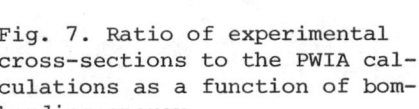
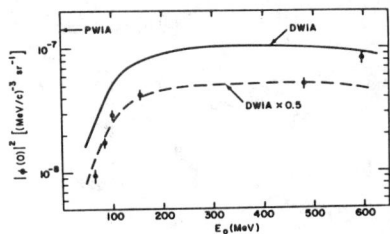

Fig. 7. Ratio of experimental cross-sections to the PWIA calculations as a function of bombarding energy.

Fig. 8. Ratio of $\sigma(\theta_1,\theta_2,E_1)/\text{PSF} \times \sigma_{pp}(\theta)$ as a function of bombarding energy for $^4\text{He}(p,2p)$. The full curve is a DWIA calculation.

the proton single particle wave function of Lim[16]. This wave function is of the Eckart form with parameters chosen to fit electron scattering, including the minimum observed at high momentum transfer. In PWIA the Lim wave function yields a cross section about a factor of two smaller than the Gaussian wave function used in the previous analysis, and thereby raises the ^{4}He curve. The optical potentials were obtained from fits to elastic scattering at lower energies, and extended to higher energies using the impulse approximation.

Fig. 8 shows the results at various energies. Here I present $\sigma(\theta_1,\theta_2,E_1)$ / PSF x $\sigma_{pp}(\theta)$ for the equal angle zero recoil momentum point. The solid curve represents the DWIA calculation, and the dashed curve is DWIA x 1/2, or if you like, assuming a proton spectroscopic factor of 1. Unfortunately the conclusions are not obvious. Clearly there is a discrepancy between the 480 MeV and 600 MeV data, and the experiment should be repeated in this energy range. If the 600 MeV data is correct, then the wave function of Lim is probably reasonable, and DWIA does not work at lower energies. If the 480 MeV data is correct, then the shape predicted by the DWIA is quite reasonable although interesting detailed discrepancies arise at the lowest energies (see the talk of E. F. Redish in this same meeting). Thus perhaps we are showing the inadequacy of the single particle wave function, thereby gaining information on the single particle wave function in ^{4}He. I feel that a good, detailed higher energy study of the ^{4}He(p,2p)t reaction is important, and has the possibility of providing very good information on the ^{4}He single particle wave function. Let me explicitly point out the interesting energy dependence predicted by the DWIA. Assuming applicability of the DWIA we find little to be gained in terms of absorption by increasing the bombarding energy beyond 100 MeV to 150 MeV. I will return to this point when I discuss distortion effects in general.

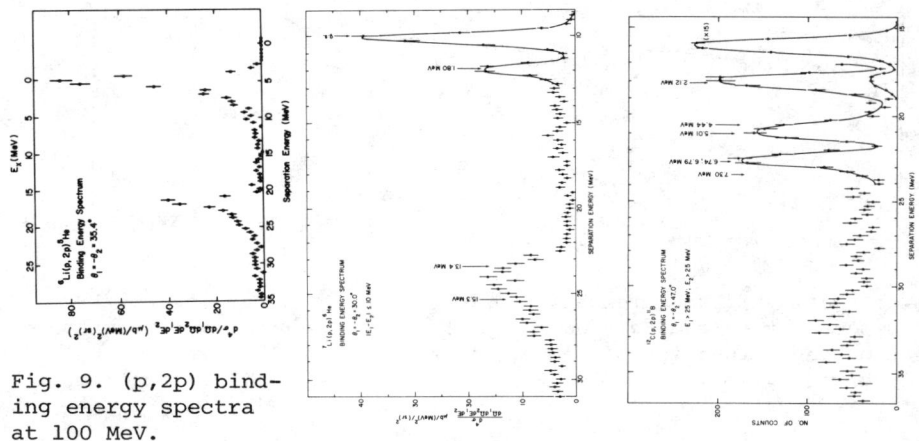

Fig. 9. (p,2p) binding energy spectra at 100 MeV.

A = 6,7,12

At Maryland we have also made detailed investigations[17] of the (p,2p) reaction at 100 MeV on ^{6}Li, ^{7}Li, and ^{12}C. In Fig. 9 I present the binding energy spectra for these three nuclei. The experimental resolution is only about 1 MeV due to the use of NaI detectors. The low lying states arise principally from p-shell proton knockout, although the 5/2 - and 7/2 - states in ^{11}C are clearly excited - presumably by a 2-step process. At high excitation energies we see the contribution from the s-state proton knockout. This stands out very clearly for ^{6}Li and ^{7}Li, but is barely visible above a rather constant background in ^{12}C.

Discussing first the p-state knockout, I want to give special consideration to ^{6}Li(p,2p)^{5}He which is very special since the transition goes to an unbound state. Thus although there is a resonance in the $p_{3/2}$ n-α channel (^{5}He "ground state"), there are also $S_{1/2}$ and $P_{1/2}$ components in the n-α system. Therefore the ^{5}He ground state region is really a mixed angular momentum "state." If in addition the ^{6}Li ground state is not pure $(1S)^4(1p)^2$ but contains configurations such as $(1S)^4(2S)^2$ as it would assuming an α + d cluster structure, we can have a mixed angular momentum transition. Thus we cannot have just L = 1 transitions, but also L = 0. Near p_3 = 0 the L = 0 cross section is much larger than the L = 1 cross section and thus small components of L = 0 will be visible. To make a long story short, this does happen and has led to much discussion over the years. We[18,19] have analyzed our ^{6}Li data basically following the procedure of Saito, et al[20] but including distortion effects by means of the DWIA as well as a better n-α wave function. In particular the n-α unbound final state for each $E_{n-α}$ was obtained from Woods-Saxon potentials which reproduce the n-α phase shifts, including the resonances. An α + d cluster model was used for the ^{6}Li ground state, thereby introducing the configuration mixing.

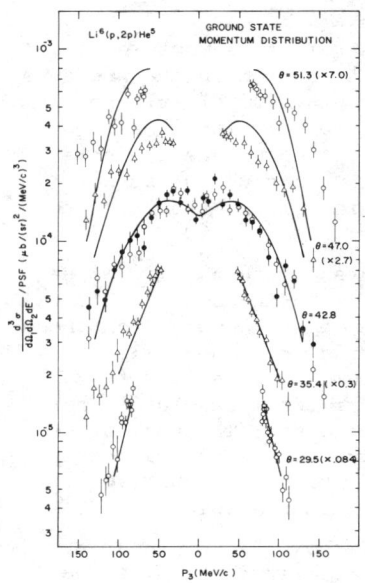
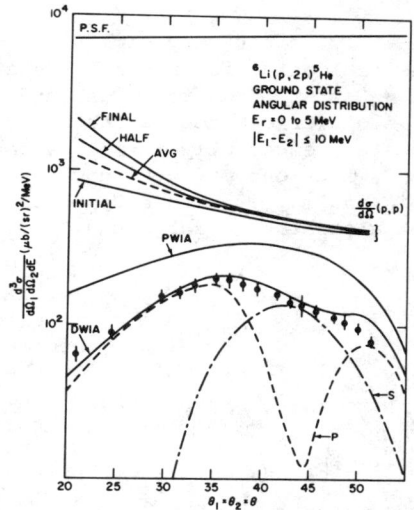

Fig. 10. Experimental momentum distributions for ^{6}Li (p,2p) at 100 MeV for equal angles. The curves are DWIA calculations.

Fig. 11. Coplanar symmetric angular distribution for ^{6}Li(p,2p) at 100 MeV. The curves are DWIA (and PWIA) for the two possible angular momenta. The upper curves show the 2-body cross section for various prescriptions.

Fig. 10 shows the energy sharing spectra for a number of equal angle pairs. Here we have plotted $\sigma(\theta_1,\theta_2,E_1)$ / PSF versus the recoil momentum p_3. The curves are the DWIA calculations with a constant normalization. At the quasi-free angle we see that there is only a very weak minimum, and that this is well reproduced by the calculation which includes the s-wave contribution to fill in the minimum. I should emphasize that the lack of a minimum is not due to experimental angular and energy resolution. At forward angles we see that the slope of the data increases and this effect is included in the DWIA. At the backward angles the distribution flattens out and the calculations do not provide quite as good agreement.

Taking the minimum recoil momentum point for each angle pair we obtain the coplanar symmetric angular distribution which is shown in Fig. 11. The curves at the top represent the various prescriptions for $\sigma_{pp}(\theta)$ and we see that the differences are not too large for this case which is not far off-shell. Again I present the DWIA calculation and this time show the s- and p- knockout contributions separately. I think this figure speaks for itself. The fit is excellent, and clearly shows the necessity for including the s-state contribution. These data and calculations are probably the best evidence for configuration mixing, or the need for an α + d cluster model, in the ^{6}Li ground state. This use of the unbound state provides a very powerful technique, but unfortunately is not very common. The measurement is most sensitive when one is looking for a small s-state contribution in the presence of an L ≠ 0 transition to an unbound state. It also clearly points to possible difficulties in the analysis of (p,2p) reactions to unbound states, especially if only a couple of final state channels are open. Thus one may get knockout contributions which are not isotropic and may lead to erroneous conclusions from attempting to subtract a constant "background."

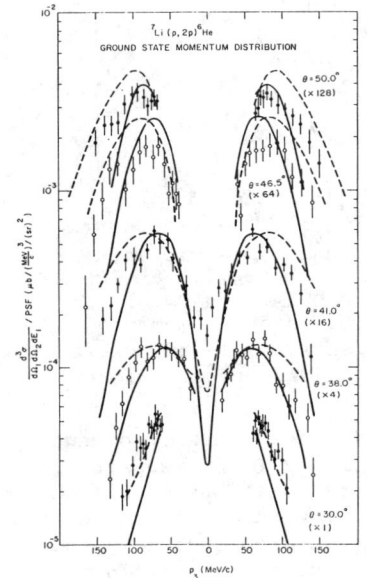

Fig. 12. Energy sharing spectra for ^{7}Li(p,2p) at 100 MeV. The curves are DWIA calculations.

Fig. 12 shows the ^{7}Li(p,2p)^{6}He (g.s.) energy sharing distribution.[21] In this case we see a very pronounced minimum showing that the filling in of the minimum in ^{6}Li arises from neither experimental resolution nor distortion effects. The two curves are DWIA calculations for different proton bound state wave functions. Again the fit is reasonably good.

Fig. 13 shows the coplanar symmetric angular distribution.[21] We see that the fit is reasonably good although it deteriorates some at forward angles, depending upon the prescription for $\sigma_{pp}(\theta)$. The spectroscopic factor is 0.24, but this can be increased by changing the bound state parameters with no real deterioration in the fit.

Finally for p-state knockout I present the results[21] for the ^{12}C(p,2p)^{11}B(g.s.) reaction. Fig. 14 shows the coplanar symmetric angular distribution and the DWIA calculation with $C^2S \simeq 2.8$ in good agreement with shell model predictions. The fit beyond the minimum is extremely good, but as in the case of ^{7}Li it deteriorates at forward angles although it is worse in this case.

I believe that this deterioration at forward angles arises from a breakdown in the factorization approximation at forward angles at 100 MeV. In order to examine this possibility we carried out DWIA and DWTA calculations[17] with McCarthy's code and pseudo-potential. The results for ^{7}Li and ^{12}C are presented in Fig. 15. We see that at least for this pseudo potential, the breakdown in factorization is primarily at forward angles and is in such a direction as to improve agreement with the experimental results. Thus at larger angles the factorization is probably quite good, but one should be very careful of the forward angle data. I should also mention that

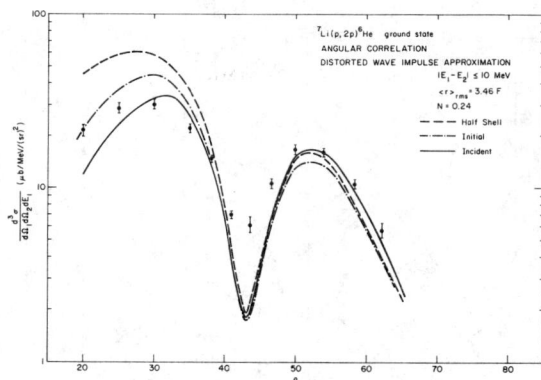

Fig. 13. Coplanar symmetric angular distribution for ^{7}Li(p,2p) at 100 MeV. The curves show a DWIA calculation for various 2-body cross section prescriptions.

Fig. 14. Coplanar symmetric angular distribution for ^{12}C(p,2p) at 100 MeV. The curves are DWIA calculations.

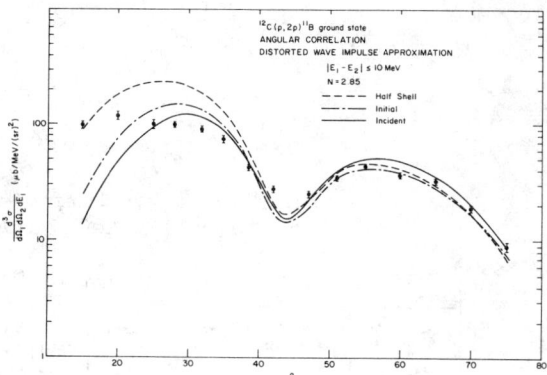

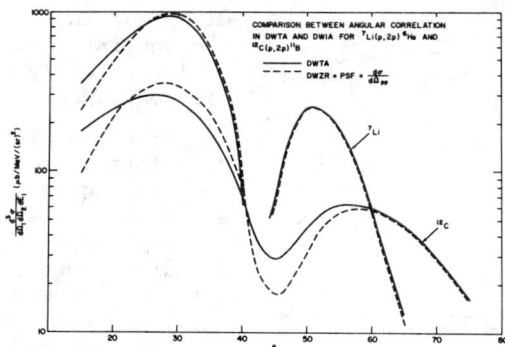

Fig. 15. DWIA (dashed line) and DWTA (full line) calculations for (p,2p) in ^{7}Li and ^{12}C at 100 MeV.

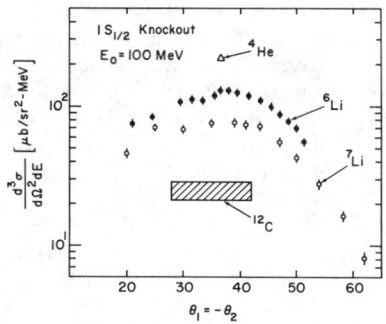

Fig. 16. Coplanar symmetric angular distribution data for s-state knockout from ^{4}He, ^{6}Li, ^{7}Li, and ^{12}C at 100 MeV.

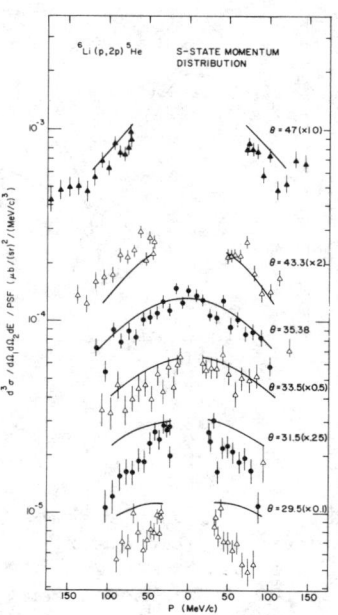

Fig. 17. Energy sharing spectra for s-state knockout from ^{6}Li at 100 MeV. The curves are DWIA calculations.

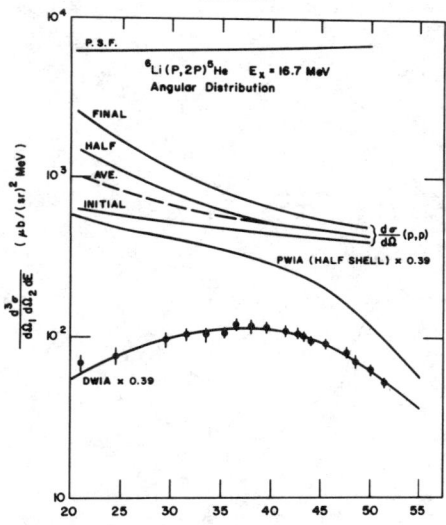

Fig. 18. Coplanar symmetric angular distribution for s-state knockout from ^{6}Li at 100 MeV. The curves are DWIA and PWIA calculations.

the shape of the calculation at these same forward angles is most sensitive to changes in the distorting potentials. Thus the forward angle data taken in the coplanar-symmetric geometry show many sensitivities, and thoughts of extracting off-shell nucleon-nucleon information from this region probably should be considered pipedreams.

For s-state knockout we have integrated over the broad peak at high excitation energies shown in Fig. 9. Fig. 16 shows the available data in the coplanar symmetric geometry. The data for ^{12}C was not systematically analyzed, but has a nearly constant magnitude of $\sim$25 μb/sr^2 MeV over the angular range indicated. I have also included the cross section for ^{4}He(p,2p)t which is basically the core for these other nuclei. These data show quite a nice systematic trend and are just about what one would expect due to an increase in distortion effects as the mass of the target is increased. Unfortunately, we have not performed a systematic DWIA calculation for these data. However, the calculations which have been done very nearly reproduce this behavior.

Fig. 17 shows some energy sharing spectra for the ^{6}Li s-state knockout. The curves are DWIA calculations. In this case we have the same trend as we had for p-state knockout; viz., at forward angles the data is steeper. Unfortunately the DWIA calculations do not reproduce this trend. However, at larger angles the agreement is reasonably good. Fig. 18 shows the coplanar symmetric angular distribution for this state. The curves above show the 2-body cross section $\sigma_{pp}(\theta)$ for the various prescriptions. Here we see that at the forward angles the differences are quite pronounced. The curve through the data is a DWIA calculation normalized to the data. The normalization factor is almost the same as that used for ^{4}He (within 20%). The DWIA calculation uses the 1/2-shell cross section and our initial reaction to the quality of the fit was exultation over the necessity of using this $\sigma_{pp}(\theta)$. However, after our later experiences with the sensitivity of the forward angles to the factorization approximation and the optical potentials we were suitably deflated. Thus this data does not prove the necessity of employing the 1/2-shell cross section.

GENERAL DISTORTION EFFECTS

In an attempt to study general distortion effects in (p,2p) reactions we have carried out a number of DWIA calculations at various energies using the code of N. S. Chant.[5] In particular we have calculated ^{12}C(p,2p) for $1P_{3/2}$ and $1S_{1/2}$ knockout, ^{40}Ca(p,2p) for $2S_{1/2}$ and $1S_{1/2}$ knockout, and ^{208}Pb(p,2p) for $3S_{1/2}$ knockout. The proton bound state Woods-Saxon well parameters were taken from Elton and Swifts' shell model analysis.[22] The optical model potentials for the entrance and exit channels were painfully obtained. For ^{12}C we took the geometrical parameters from a 61 MeV p-scattering analysis of Fulmer, et al[23]. The well depths as a function of energy were chosen to smoothly reproduce the volume integrals (real and imaginary) of analyses[24] of data from 30 MeV to 1 GeV. The same procedure was followed for ^{40}Ca and ^{208}Pb using the optical model analyses of Refs. 25 and 26. Thus the potentials have a nice systematic variation with energy. Where possible, or worthwhile, calculations were carried out at 50, 100, 150, 200, 300, and 400 MeV.

Fig. 19 shows the ratio of the DWIA calculation to the PWIA calculation as a function of energy. The calculations correspond only to equal angle quasi-free angle pairs at zero recoil momentum. We see the same trend as I showed in the case of ^{4}He; viz., there is a rapid increase at lower energies, and then the ratio flattens. Thus above a certain energy, the distortion effects remain approximately constant. The details of course depend on the binding energy and the A of the nucleus.

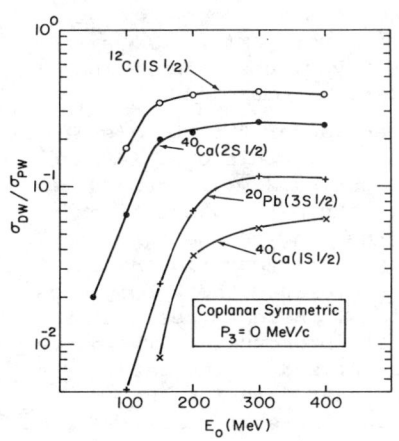

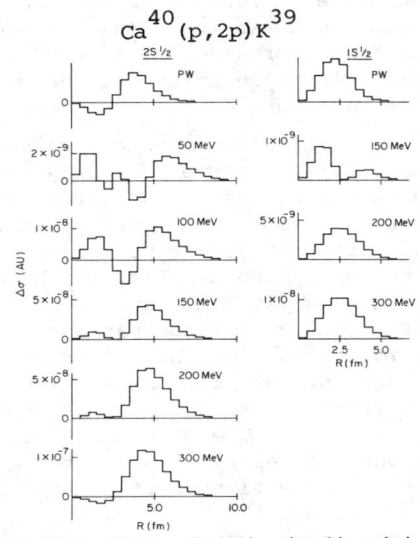

Fig. 19. The ratio $\sigma_{DWIA}/\sigma_{PWIA}$ as a function of bombarding energy for (p,2p) knockout of S-state protons.

Fig. 20. Plot of $\Delta\sigma (=\sigma(R+\Delta)-\sigma(R))$ versus R for 1S and 2S knockout from ^{40}Ca.

Based on this display one would argue that from a spectroscopy standpoint a bombarding energy of the order of 150 to 200 MeV is adequate for most (p,2p) studies and there is no need to go to a higher energy where the experiments are generally more difficult. From a reaction mechanism standpoint I think it would be interesting to study the (p,2p) reaction from 50 to 200 MeV, to see if the DWIA reproduces this behavior.

In an attempt to understand this behavior and to understand the region of sensitivity of the (p,2p) reaction, we have calculated the (p,2p) cross section contribution as a function of cutoff radius. The code prints out this information for each calculation, so no additional calculations are required. In Fig. 20 we have plotted the quantity $\Delta\sigma = \sigma$ (R cutoff = r + 0.5) - σ (R cutoff = r) as a function of radius. This plot should be at least an indication of the sensitivity of the reaction to various radial regions. The calculations correspond to zero recoil momentum for the equal angle quasi-free angle pair for the $2S_{1/2}$ and $1S_{1/2}$ states of ^{40}Ca. The top distribution corresponds to PWIA with no distortion effects. Let me note the following. First there is a large change in scale between the lower and higher energies. Secondly, at lower energies in addition to the reduced contribution, there are clearly pronounced phasing effects. This presumably arises from the importance of the real potential at these lower energies - the real potential is comparable to the energies of the particles. Thirdly, at the higher energies the $\Delta\sigma$ begins to rather closely resemble the PWIA calculation. At these energies the real well is small compared to the proton energy, and the distortion effects are dominated by absorption. Thus we just have a gradual, but important, reduction as we penetrate into the nucleus. Finally, no matter what the energy the nuclear interior plays a significant role in some fashion.

The second study we made was to carry out detailed calculations at 100 MeV and 200 MeV for both the coplanar symmetric angular distribution (CSAD) and the energy sharing (ES) distribution at the equal angle quasi-free angle pair. The calculations were carried out for an angular range (CSAD) of roughly 15° to 60°, and an energy range (ES) from about 20 MeV to the equal energy point. In order to facilitate the comparisons I have plotted the calculated distorted momentum distributions ϕ^2 versus recoil momentum p_3. For the energy sharing distribution at equal angles the distribution must be symmetric about $p_3 = 0$ due to the identity of the two protons. For the CSAD I have plotted the forward angles to the left of $p_3 = 0$ and the large angles to the right of $p_3 = 0$.

Fig. 21 shows the calculated ϕ^2 for $P_{3/2}$ knockout from ^{12}C. The dashed curve is the PWIA calculation (i.e., the square of the Fourier transform of the wave function) normalized by the factor indicated. At 100 MeV we see large differences between DWIA and PWIA, and we see that the minimum in the CSAD is shifted a great deal. In neither case can one say that the ϕ^2_{DW} reflects the true momentum distribution. However, at 200 MeV the ES calculation is strikingly similar to the PWIA ϕ^2. The CSAD is not nearly as good being too low at forward angles and too high at large angles.

Fig. 22 shows the $2S_{1/2}$ knockout from ^{40}Ca at 100 MeV. Here the ES distribution is most reasonable, but shows no evidence of the nodes in the wave function. Fig. 23 shows the same state at 200 MeV. In this case we see the very lovely feature of nodes (or at least minima) in the ES distribution. In fact the ES distribution reflects most of the important features of the distribution. The CSAD on the other hand would lead to extremely misleading results,

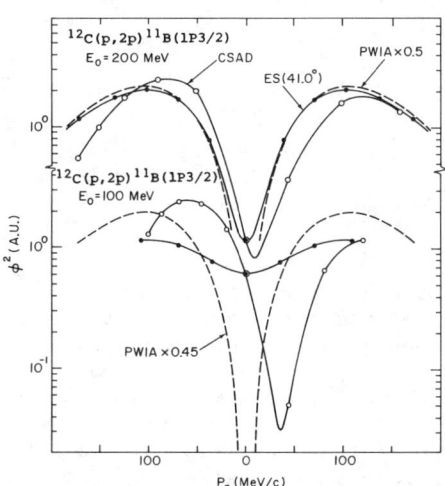

Fig.21. Calculated momentum distributions. DWIA results are presented for energy sharing (ES) and coplanar symmetric (CSAD) experimental arrangements.

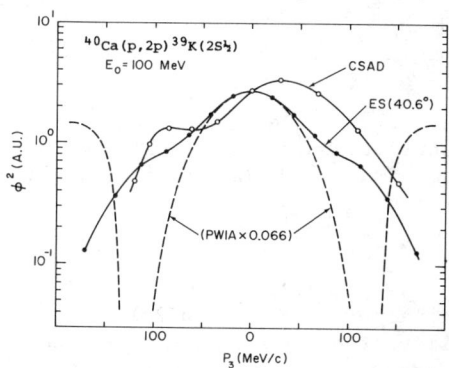

Fig.22. Calculated momentum distributions for $2S_{1/2}$ knockout from ^{40}Ca. (See Fig.21)

if interpreted in a plane wave formalism.

The final figure (Fig. 24) shows the $3S_{1/2}$ knockout from ^{208}Pb at 200 MeV. Again the results are similar to those I presented previously.

I think that the conclusions from this array of figures is obvious. In terms of spectroscopy one wants to perform the ES experiment, not the CSAD, around 200 MeV. Doing this experiment first, one can obtain information on the wave function. Then, one can carry out the CSAD in order to study the reaction mechanism, distortion effects, factorization, and/or off-shell effects. I might also add that I expect the ES experiment to be less sensitive to a factorization breakdown, since one is really using the two-body cross section over a relatively small range of energy and angle, and at a large enough angle where the breakdown does not appear to be too great. It is unfortunate that almost all experiments have concentrated on the CSAD*, when in fact the simpler ES experiment should have been carried out first.

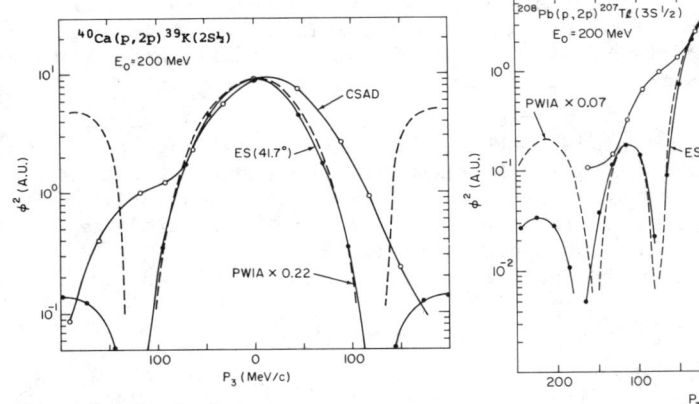

Fig.23. Calculated momentum distributions for $2S_{\frac{1}{2}}$ knockout from ^{40}Ca at 200 MeV (see Fig. 21).

Fig.24. Calculated momentum distributions for $3S_{\frac{1}{2}}$ knockout from ^{208}Pb at 200 MeV (see Fig. 21).

CONCLUSIONS

Very briefly I hope that I have shown a number of the interesting aspects of the (p,2p) reaction. Over the past few years I believe that too little attention has been given to these reactions. A careful, systematic investigation of the (p,2p) reaction which can now be performed with high precision accelerators in the 100 to

---

* Professor Maris pointed out at this point that the original reason for choosing the CSAD was to remove polarization effects, since the p-p scattering angle is always 90°. This is an important consideration. Recent polarized (p,2p) studies at TRIUMF[27] should indicate the importance of polarization effects in the ES distribution.

400 MeV range should provide quantitative nuclear structure information previously unattainable.

ACKNOWLEDGEMENTS

I would like to acknowledge all of my colleagues at Maryland, particularly Dr. N. S. Chant, for many useful and stimulating discussions.

REFERENCES

1. E. F. Redish, G. J. Stephenson and G. M. Lerner, Phys. Rev. $\underline{C2}$ 1665 (1970).
2. R. Aaron, R. D. Amado and Y. Y. Yam, Phys. Rev. $\underline{136}$ B650 (1964); Phys. Rev. $\underline{140}$ B1291 (1965); R. Aaron and R. D. Amado, Phys. Rev. $\underline{150}$ 857 (1970).
3. G. Jacob and T. A. J. Maris, Rev. Mod. Phys. $\underline{38}$ 121 (1966); Rev. Mod. Phys. $\underline{45}$ 6 (1973).
4. D. F. Jackson and T. Berggren, Nucl. Phys. $\underline{62}$ 353 (1965).
5. N. S. Chant and P. G. Roos, Proc. of Second Int. Conf. on Clustering Phenomena in Nuclei, College Park, Md., 1975, ed. by D. A. Goldberg, J. B. Marion, and S. J. Wallace, ORO-4856-26 (ERDA Tech. Info. Center, Oak Ridge, Tennessee, 1975) p. 265.
6. K. L. Lim and I. E. McCarthy, Nucl. Phys. $\underline{88}$ 433 (1966).
7. Richard D. Koshel, Nucl. Phys. $\underline{A260}$ 401 (1976); R. D. Koshel, this conference.
8. P. G. Roos and N. S. Chant, Proc. of Second Int. Conf. on Clustering Phenomena in Nuclei, College Park, Md., 1975, ed. by D. A. Goldberg, J. B. Marion, and S. J. Wallace, ORO-4856-26 (ERDA Tech. Info. Center, Oak Ridge, Tennessee, 1975) p.242.
9. H. G. Pugh, et al, Phys. Rev. Lett. $\underline{22}$ 408 (1969); J. W. Watson, et al, Nucl. Phys. $\underline{A172}$ 513 (1971).
10. P. G. Roos, H. Kim, M. Jain, and H. D. Holmgren, Phys. Rev. Lett. $\underline{22}$ 242 (1969).
11. J. M. Wallace, Phys. Rev. $\underline{C7}$ 10 (1973); Phys. Rev. $\underline{C8}$ 1275 (1973).
12. V. K. C. Cheng and P. G. Roos, Nucl. Phys. $\underline{A225}$ 397 (1974); V. K. C. Cheng, Ph.D. Thesis, Univ. of Md. (1973).
13. H. G. Pugh, et al, Phys. Lett. $\underline{46B}$ 192 (1973).
14. R. D. Haracz and T. K. Lim, Phys. Rev. Lett. $\underline{31}$ 1263 (1973).
15. P. G. Roos, Phys. Rev. $\underline{C9}$ 2437 (1974); R. Frascaria, et al, Phys. Rev. $\underline{C12}$ 243 (1975).
16. T. K. Lim, Phys. Lett. $\underline{44B}$ 341 (1973).
17. R. K. Bhowmik, Ph.D. Thesis, Univ. of Md. (1974).
18. R. K. Bhowmik, C. C. Chang, P. G. Roos and H. D. Holmgren, Nucl. Phys. $\underline{A226}$ 365 (1974).
19. C. C. Chang, R. K. Bhowmik, N. S. Chant and P. G. Roos, submitted to Nucl. Phys.
20. S. Saito, J. Hiura and H. Tanaka, Prog. Theor. Phys. $\underline{39}$ 635 (1968).
21. R. K. Bhowmik, C. C. Chang, J. P. Didelez and H. D. Holmgren, Phys. Rev. C, to be published.
22. L. R. B. Elton and A. Swift, Nucl. Phys. $\underline{A94}$ 52 (1967).

23. C. B. Fulmer, J. B. Ball, A. Scott and M. L. Whiten, Phys. Rev. 181 1565 (1969).
24. G. Igo, private communication.
25. W. T. H. van Oers, Phys. Rev. C3 1550 (1971).
26. W. T. H. van Oers, et al, Phys. Rev. C10 307 (1974).
27. D. Hutcheon, private communication.

INVESTIGATION OF ATOMIC AND MOLECULAR ELECTRONIC STRUCTURE BY USE OF THE (e,e') REACTION

R. A. Bonham[*]
Indiana University, Bloomington, Ind. 47401

ABSTRACT

A survey of electron impact spectroscopy in the 20 keV to 40 keV range is presented. Work on the measurement of Bethe surfaces, charge densities, Compton profiles and inner and outer shell spectroscopy is reviewed.

INTRODUCTION

This paper will attempt to present a number of experimental realizations of the Bethe-Born theory which was so ably reviewed by Inokuti.[1] Most of the examples to be presented were obtained from the Indiana University high energy electron impact spectroscopy (HEEIS) laboratory.

The basic experiment involves an incident electron beam eminating from a telefocus type electron gun[2] with beam current ranging from 10^{-3} μA to 5×10^2 μA with incident energies adjustable between 20 keV and about 50 keV. The beam is focused on the entrance slit of the energy analyzer about 1 m from the gun with an FWHM of about 200 μm. The target is a gas jet produced by expansion through a 0.125 mm ID hypodermic needle of 6 mm length. The local gas density in the jet is $10^{-3} - 10^{-1}$ Torr near the nozzle tip[3] while a low vacuum pressure is maintained nearby ($\sim 10^{-5}$ Torr). The basic experimental configuration is shown in Fig. 1. Note that the energy

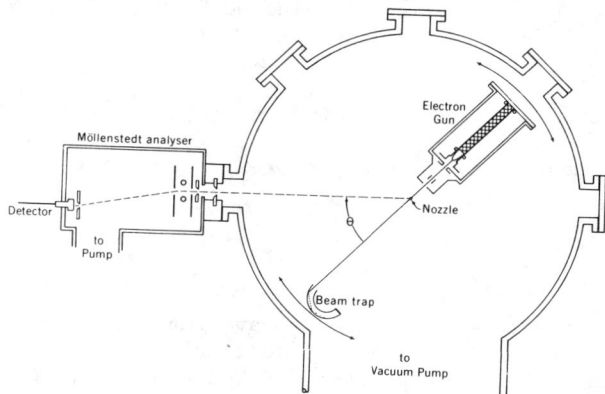

Fig. 1. A cutaway view of the high-energy electron spectrometer

[*]The author wishes to acknowledge support by the National Science Foundation, Grant No. GP-41983X and to the Donors of the Petroleum Research Fund, administered by the American Chemical Society.

analyzer is connected to the scattering chamber[3] whose pumping system has exhibited measured pumping speeds for He of 20,000 ℓ/sec and for N_2 of 13,000 ℓ/sec. The electron gun is mounted in such a way that it can be rotated about the scattering center so that the scattered electrons can be analyzed for energy transferred to the target (energy loss) as a function of the scattering angle. The energy analyzer is of the Mollenstedt type[4] with a resolution capability of less than 100 meV over an energy loss range of at least 4,000 eV. Actual energy resolution is limited by the energy spread in the incident electron beam which is typically 500 meV for the tungsten filaments most commonly used. In Table 1 a summary of the relevant experimental parameters is presented.

THEORY

It appears worthwhile to present a very quick review of the Bethe-Born theory presented yesterday by Dr. Inokuti. The cross section for high energy electron inelastic scattering differential with respect to both energy loss, E, and the solid angle of observation, Ω, is given in the Born approximation by

$$\frac{d^2\sigma}{d\Omega dE} = \left(\frac{d\sigma}{d\Omega}\right)_{Rutherford} \left[\frac{k(E)}{kE} K^2\right] \frac{df(K,E)}{dE} \qquad (1)$$

where $\left(\frac{d\sigma}{d\Omega}\right)_{Rutherford} = 4/K^4$ in Rydberg atomic units, K is the momentum transferred to the target, $k(E) = \sqrt{k^2 - E}$ is the wave vector for the scattered electron and $df(K,E)/dE$ is the generalized oscillator strength (GOS) per unit energy loss. Note that projectile electron exchange with target electrons is neglected in Eq. 1 but may be approximately included by replacing

$$\left(\frac{d\sigma}{d\Omega}\right)_{Rutherford} \text{ by } \left(\frac{d\sigma}{d\Omega}\right)_{Mott} \cong \left(\frac{d\sigma}{d\Omega}\right)_{Rutherford} \left[1 - \frac{K^2}{k^2} + \frac{K^4}{k^4}\right].$$

The GOS is the quantity of central interest and is given by

$$\frac{df(K,E)}{dE} = \sum_n E_n |\langle \Psi_n | \sum_{i=1}^{N} e^{i\vec{K}\cdot\vec{r}_i} | \Psi_0 \rangle|^2 \delta(E-E_n)/K^2 \qquad (2)$$

where Ψ_0, Ψ_n are the target electronic wave functions for the initial and final states of the target and $\vec{r}_i$ defines the instantaneous position of the ith of N target electrons. The sum in Eq. 2 is limited to all possible excitation processes accessible to the incident projectile energy k^2 with energy loss E. In the case of a bound state excitation only one term in the sum in Eq. 2 exists for a nondegenerate state and in that case the notation $f(K,E)$ is normally used.

In the non physical limit of zero momentum transfer the GOS reduces to

TABLE I

EXPERIMENTAL PARAMETERS

Electron Beam Source
- Incident Electron Energy $20 \text{ keV} < E_0 < 45 \text{ keV}$
- Beam Current $0.01 \text{ }\mu\text{Amp} < I < 500 \text{ }\mu\text{Amp}$
- Focused Beam Size $180 \text{ }\mu\text{m}$ FWHM
- Beam Size at Target $250 \text{ }\mu\text{m}$ FWHM
- Energy Spread in Incident Beam $\sim 0.5 \text{ eV}$

Target Gas Jet
- Flow Rate (F) $10^{16} \text{ molecules/sec} < F < 10^{20}$
- Jet Diameter (1 mm from Tip) $\sim 1 \text{ mm}$ FWHM
- Target Pressure $10^{-3} \text{ Torr} < P_T < .3 \text{ Torr}$
- Background Pressure $10^{-6} < P_B < 10^{-4}$

Energy Analyzer
- Solid Acceptance Angle $10^{-8} \text{ SR} < d\Omega < 10^{-5} \text{ SR}$
- Energy Loss Range $0 \leq E < 4 \text{ keV}$
- Resolution $\sim 0.1 \text{ eV}$

Scattering Angle
- Range $0° < \theta < 90°$
- Mechanical Accuracy $\Delta\theta \sim .005°$
- Accuracy (Including Magnetic Effects) $\Delta\theta \sim .02°$

Detector   Si Surface Barrier
- Max Count Rate   50 kHz
- Lowest Electron Energy for Effective Noise Elimination   23 keV
- Dead Time $\sim \text{ }\mu\text{sec}$

$$\frac{df(E)}{dE} = \sum_n E_n |\langle \Psi_n | \sum_{i=1}^{N} z_i | \Psi_0 \rangle|^2 \delta(E-E_n) \qquad (3)$$

where $df(E)/dE$ is called the optical oscillator strength and the matrix element is the familiar dipole transition matrix element encountered in photoabsorption spectroscopy. In fact the photon absorption cross section in Rydberg atomic units can be written as[5]

$$\frac{d\sigma}{dE} = \frac{\pi}{c} \frac{df(E)}{dE} \qquad (4)$$

The corresponding electron case is given by

$$\frac{d^2\sigma}{d\Omega dE} = \frac{16k(E)k}{E^3} \frac{df(E)}{dE} \qquad (5)$$

which establishes the close connection between the photoabsorption intensity and the high energy electron impact inelastic scattering in the forward direction. Prof. Lassettre[6] has argued that if one can reliably extrapolate small angle intensity data to the $K = 0$ limit, note that the physical limit $\theta = 0°$ is approximately $K=E/2k$, then the correct value of $\frac{df(E)}{dE}$ must be obtained independently of whether or not the Born approximation holds.

If the cross section given in Eq. 1 is relatively constant over a small range of angles near $\theta = 0°$ with the solid angle of detection less than this range then the ratio of photon absorption to electron scattering is approximately

$$\frac{(\frac{d\sigma}{dE})_P}{(\frac{d\sigma}{dE})_e} \simeq \frac{\pi E^3}{16ck(E)kd\Omega} \qquad (6)$$

which is independent of the nature of the target. For incident electron energies around 40 keV and solid angles of detection of $10^{-6} - 10^{-8}$ rad^2 the ratio in Eq. 6 indicates that the photon cross section may be larger than the electron one by a factor of about 10^4 at an energy loss of 1,000 eV. The situation below 1 keV changes in favor of the electron method as one goes to low energy loss. However, it is a simple matter to build electron beam sources by a factor of at least 10^4. This means that the electron impact method is a useful low cost alternative to photon spectroscopy for use in investigating gas, thin film and surface targets.

For angles greater than $\theta = 0°$ the comparison between electron scattering and photon scattering is even more decisively in favor of the electron method since in this case the relevant cross section ratio is given approximately by the ratio of the x-ray Thompson cross section[7] to the Rutherford or Mott cross sections. For medium momentum transfer ($K \sim 10$ a.u.) the electron impact method has a cross section advantage of at least 10^4 before source densities are even considered. Hence for investigations on Compton profiles, for example, the electron impact method has distinct advantages.

The central task in HEEIS work is to reduce measured intensity values to relative oscillator strengths and to then place these on an absolute scale. This is normally accomplished by use of Eq. 1 with the Mott cross section to define relative GOS values. These are then placed on an absolute scale by use of the zeroth moment of the GOS. The ℓth moment of the GOS can be written as

$$S(\ell,K) = \int_0^\infty dE \, E^\ell \frac{df(K,E)}{dE} \tag{7}$$

which for $\ell = 0$ yields zeroth moment or the Bethe sum rule[1]

$$S(0,K) = N \tag{8}$$

where N is the number of electrons in the target system. The sum rule for $\ell = -1$ is also of interest in connection with charge density measurements since

$$S(-1,K) = \frac{S(K)}{K^2} \tag{9}$$

where the x-ray incoherent scattering factor, $S(K)$, can be written as

$$S(K) = N + \int_0^\infty dr \, P(r) j_0(Kr) - F(K)^2 \tag{10}$$

where $P(r)$ is the electron pair correlation function,[8] $j_0(Kr)$ is the zero order spherical Bessel function and $F(K)$ is the coherent x-ray scattering factor[8]

$$F(K) = \int d\vec{r} \, \rho(\vec{r}) e^{i\vec{K}\cdot\vec{r}} \tag{11}$$

with $\rho(\vec{r})$ the one electron charge density of the target. The electron pair correlation function is known to be sensitive to the effects of electron correlation.

In Fig. 2 a plot of $\frac{df(K,E)}{dE}$ vs. $\ell n[K^2]$ and E is constructed from cuts made at selected values of the scattering angle for 25 keV electrons scattered from H_2 on an absolute scale.[9] The momentum transfer K is nearly constant in E for fixed θ over the width of the spectrum. Such a plot is termed a Bethe surface[1] and displays complete information on possible energy exchanges between the projectile particle and the target. In Fig. 3 the scattered intensity on a relative scale, including the elastic line, is shown for 25 keV electrons scattered by N_2.[10] This spectrum is more interesting because the nitrogen atom possesses an inner shell.

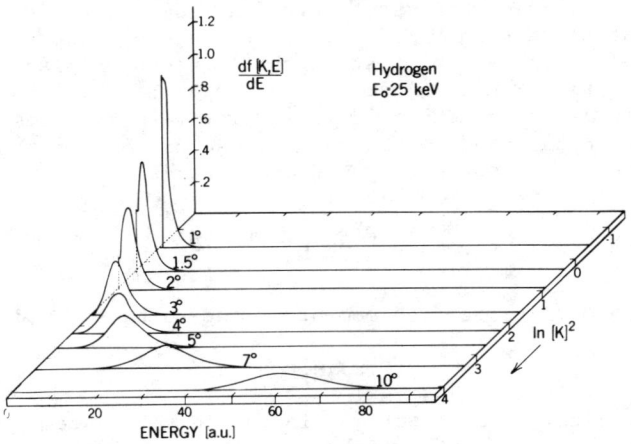

Fig. 2. The Bethe surface for H_2 as determined by 25 keV electron inelastic scattering. All scales are absolute and all units are Rydberg atomic units.

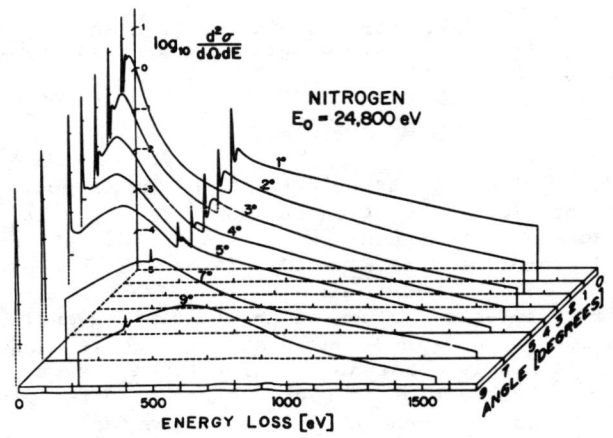

Fig. 3. The relative cross section $d^2\sigma/d\Omega dE$ as a function of scattering angle in degrees and energy loss in eV for 25 keV electrons incident on N_2.

A brief survey of Fig. 3 suffices to define the areas of interest for further detailed investigations. For example the angular dependence of the elastic line is of interest in measuring one electron charge densities. Values of S(K) derived from the inelastic scattering yield charge density information as discussed earlier. The dominant ridge in the surfaces shown in both Figs. 2 and 3 whose maximum moves to higher energy loss with increasing momentum transfer can be understood as the electron analogue of x-ray Compton scattering. Hence the shape of this ridge can be related to the one electron momentum density of the target. The various sharp spectral features in Fig. 3 offer further possibilities for obtaining information on the electronic structure of the target system. The remainder of this manuscript will be devoted to a presentation of the recent results obtained in these various categories.

Before proceeding, however, it is of interest to survey the experimental uncertainties involved in the present measurements of inelastic scattering. These are shown in Table II. Note that the uncertainty in the overall oscillator strengths is for the experimental value corrected for the correctable effects enumerated in Table II. Another source of uncertainty not listed in Table II is the question of the validity of the Born approximation. In Fig. IV we compare some absolute GOS data for N_2 obtained in our laboratory by Dr. Lee[11] with earlier results at an electron impact energy of 500 eV[12] for various energy loss values in the ionization continuum and the limiting oscillator strength values obtained by photoabsorption experiments.[13] The present results were obtained at 25 keV and 35 keV and the agreement argues in favor of the validity of the Born approximation at these energies. The results are also consistent with the optical limiting values and Lassettre's results for small momentum transfers in all cases. However significant deviations are encountered both with increasing energy loss and increasing momentum transfer. These deviations are believed to be due to a failure of the Born approximation at the lower incident energy (500 eV) and illustrate the need for collecting data at several different energies as a check on the Born assumption.

CHARGE DENSITY MEASUREMENTS

For electrons elastically scattered from neutral atoms with spherically symmetric ground states the Born cross section $(\frac{d\sigma}{d\Omega})_{el}$, normalized to the Rutherford cross section is given by

$$\frac{(\frac{d\sigma}{d\Omega})_{el}}{(\frac{d\sigma}{d\Omega})_R} = Z^2 - 2Z \int_0^\infty dr\, D(r) j_0(Kr)$$

$$+ \int d\vec{r} \rho(\vec{r}) \int d\vec{r}' \rho(\vec{r}') j_0(K|\vec{r}-\vec{r}'|) \qquad (12)$$

where Z is the nuclear charge, D(r) is the one electron radial distribution and $\rho(\vec{r})$ is the one electron density. In the limit of

TABLE II

INTENSITY CORRECTIONS AND UNCERTAINTIES IN OBTAINING THE GOS

Finite Scattering	< 3% for E > 5 eV with appropriate choice of $d\Omega$
Relativistic Corrections	4-8%
Resolution Correction	~ 3%
(Energy Resolution of Analyzer is a Function of Incident Energy-Kollath)	
Dead Time	6% at 10 Khz
Detector Efficiency Correction	Assumed to be constant
Correction for Exchange Scattering	< 3% for $0 < \theta < 10°$
Uncertainty in Energy Scale	.01 a.u.
Uncertainty in Angular Scale	.02°

Overall Uncertainty in Absolute GOS for E = K = 1 a.u. and 10^4 Counts is about 4%

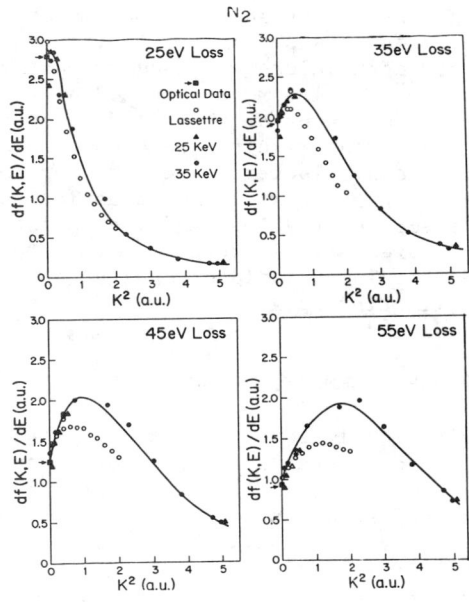

Fig. 4. Absolute oscillator strengths for N_2 at energy losses of 25, 35, 45 and 55 eV obtained with 25 keV, -Δ-Δ-Δ-, and 35 keV, -·-·-·-, electrons. Electron impact results at 500 eV are shown by the open circles[12] and the solid square, → ■ with an arrow marks the limiting oscillator strength value.[13]

large momentum transfer both integrals in Eq. 12 vanish and the Coulomb limiting value of Z^2 is approached. The inelastic scattering can be analyzed in terms of charge densities as outlined in Eqs. 10 and 11. Note the limiting value N for large momentum transfer. The sum

$$\frac{(\frac{d\sigma}{d\Omega})_{el}}{(\frac{d\sigma}{d\Omega})_R} + S(K) = Z^2 + N - 2Z\int_0^\infty drD(r)j_0(Kr) + \int_0^\infty drP(r)j_0(Kr) \qquad (13)$$

is close to the experimental result obtained by measuring the total scattering. It is important in careful work however to note that K in S(K) is an average over the possible energy losses and for large K (scattering angle) may differ from the elastic momentum transfer at the same angle.

It is convenient to define the function

$$\sigma(K) = \frac{(\frac{d\sigma}{d\Omega})_{el}}{(\frac{d\sigma}{d\Omega})_R} + S(K) - Z^2 - N \qquad (14)$$

since the area under this function has been shown by Tavard[14] to be proportional to the average electronic potential energy of the system, $\bar{V}$, as

$$\frac{2}{\pi}\int_0^\infty dK\sigma(K) = \bar{V} = \bar{V}_{en} + \bar{V}_{ee} \qquad (15)$$

where the average electron-nuclear potential energy is

$$\bar{V}_{en} = -2Z\int_0^\infty \frac{drD(r)}{r} \qquad (16)$$

and the average electron-electron potential energy is

$$\bar{V}_{ee} = \int_0^\infty \frac{drP(r)}{r} \qquad (17)$$

In Fig. 5 experimental relative total scattering results for Ne[15] are compared in a sensitive manner with theoretical results using a Hartree-Fock (HF) description of the target and a configuration interaction (CI) wave function which gives 86% of the contribution of correlation effects to the total energy (86% of the difference between the HF limit and the exact non relativistic value). Note that the experiment possesses sufficient sensitivity to clearly distinguish the small correlation effects in the small angle region.

In the case of scattering from molecules the foregoing analysis still applies except for one important modification. The normalized total scattering for a neutral molecule is now approximately given by

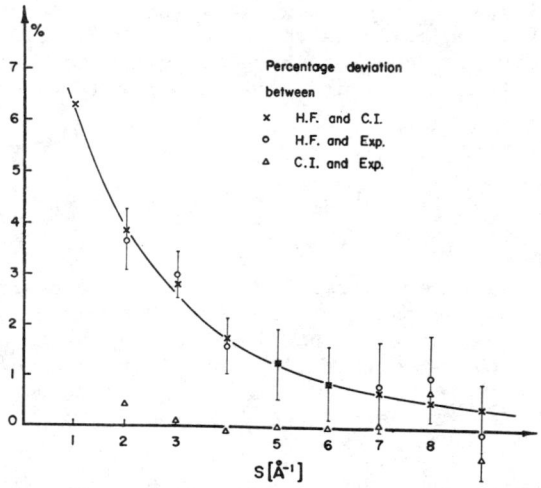

Fig. 5. Comparison between HF and CI calculated scattered intensities and HF and the experimental intensity. x: $100[I_{total}(HF)-I_{total}(CI)]/I_{total}(HF)$; O: $100[I_{total}(HF)-I_{total}(expt)]/I_{total}(HF)$; Δ: $100[I_{total}(CI)-I_{total}(expt)]/I_{total}(HF)$.

$$\frac{(\frac{d\sigma}{d\Omega})_T}{(\frac{d\sigma}{d\Omega})_R} \simeq \sum_k Z_k + N + \sum_{k \neq \ell} \sum Z_k Z_\ell j_0(KR_{k\ell})$$

$$- 2\sum_k Z_k \int d\vec{r} \rho(\vec{r}) j_0(K|\vec{r}-\vec{R}_k|)$$

$$+ \int_0^\infty dr P_0(r) j_0(Kr) \qquad (18)$$

where Z_k is the atomic number of the k<u>th</u> nucleus, $\vec{R}_k$ is the instantaneous position of the k<u>th</u> nucleus and $P_0(r)$ is the three dimensional electron pair correlation function averaged over all orientations in space. It is understood that both sides of Eq. 18 are averaged over the vibrational motion of the molecule. Also note that the molecular counterpart of Tavard's theorem[14,8] contains a new term corresponding to the nuclear-nuclear Coulomb repulsive potential energy. The main difference however is found in the electron-nuclear interaction term which now depends on all radial components of the expansion of the three dimensional (3D) one electron charge density in spherical harmonics. This situation suggests the possibility of recovering the 3D density from one dimensional scattering information. Of course such a transformation is non unique but in certain special cases where the expansion involves only 2 or 3 important terms information of this type may be possible to obtain.

From Tavard's theorem it is clear that if one subtracts the total scattering fro the separated atoms from the molecular scattering the area under the resulting difference curve is by definition the chemical binding energy of the molecule. This observation suggests that we form molecular intensity functions by subtracting the theoretically predicted atomic scattering (a relatively easy task these days) from the experimental molecular intensity result. These functions should exhibit the fine details of charge rearrangement in going from the atomic model to the real molecule. So far only the total scattering has been discussed. It is also possible to obtain additional information by looking at the elastic and inelastic scattering separately. So far this has been done experimentally only for H_2.[9] The results for the elastic comparison function is shown in Fig. 6 where the solid line marked B corresponds to a theoretical prediction using an almost exact wave function. In Fig. 7 theory and experiment for the inelastic difference function are shown. Clearly further refinements in the experiment are required. An important question is that with the rapid improvements in computational techniques is experimental work of this type really justified? So far most work has dealt with simple systems where the theory can be regarded as a calibration of the experimental method. Note that there are increasing numbers of integrals as the system becomes more complex so that the theoretical approach is not yet in a position to make apriori predictions for complex cases. Once the experiment is calibrated the difficulty in obtaining the experimental data is almost independent of the complexity of the scatterer.

Recently Prof. Fink and coworkers[16] at the University of Texas have completed a new generation scattering experiment with greatly increased accuracy. They were able to obtain new data on N_2 shown in Fig. 8. Because N_2 is a rather simple molecule, it was possible to obtain a three dimensional electron density difference map directly from the experiment with a minimum number of assumptions. Their difference map is shown in Fig. 9. Because the elastic and inelastic scattering were not separated in this experiment an assumption had to be made concerning the nature of the function $P_0(r)$ (Eq. 18) in order to extract the one electron density map. In Fig. 10 the theoretically predicted map at the HF limit is shown. The differences are clearly significant. However it is possible that these differences are due to inadequacies in the theoretical description of the electron correlation sensitive function $P_0(r)$. Either way the new data shows every promise of providing an experimental test of computed molecular wave functions beyond the HF limit.

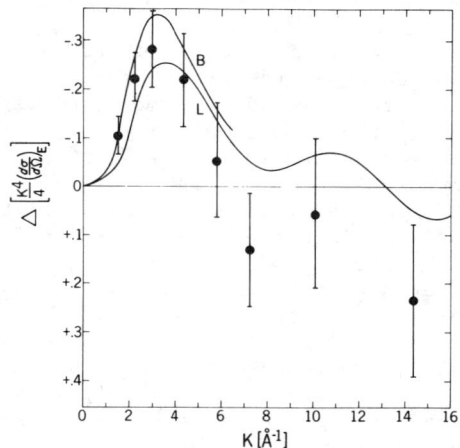

Fig. 6. Difference between the experimental elastic scattering obtained with 25 keV electrons normalized by the Rutherford cross section and the corresponding quantity predicted for two hydrogen atoms separated by the same internuclear distance as a function of momentum transfer. The data were placed on an absolute scale by use of the Bethe sum rule.

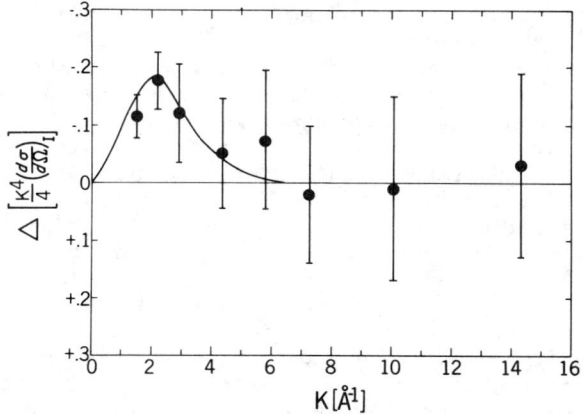

Fig. 7. Difference between the experimental total inelastic scattering obtained with 25 keV electrons normalized by the Rutherford cross section and the corresponding quantity predicted for two hydrogen atoms separated by the same internuclear separation as a function of the momentum transfer. The data were placed on an absolute scale by use of the Bethe sum rule.

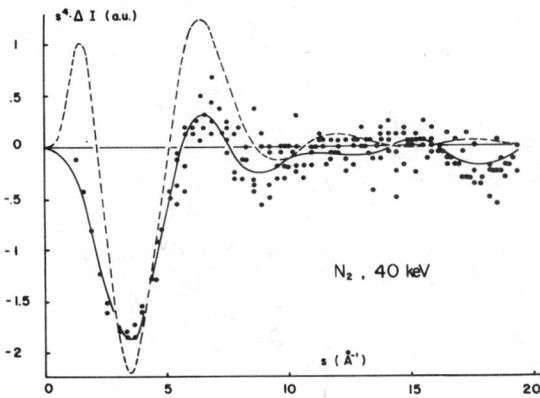

Fig. 8. Difference between the total experimental scattering for N_2 obtained with 40 keV electrons normalized by the Rutherford cross section and the same quantity predicted for spherically averaged ground state nitrogen atoms at the molecular internuclear separation as a function of momentum transfer. The data were scaled to theory at large values of the momentum transfers.

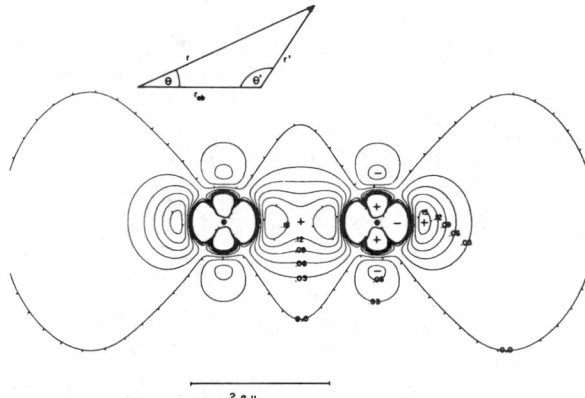

Fig. 9. Difference in the one electron density obtained from the experiment of Ref. 16 and that predicted for two spherically averaged ground state nitrogen atoms in the same nuclear configuration.

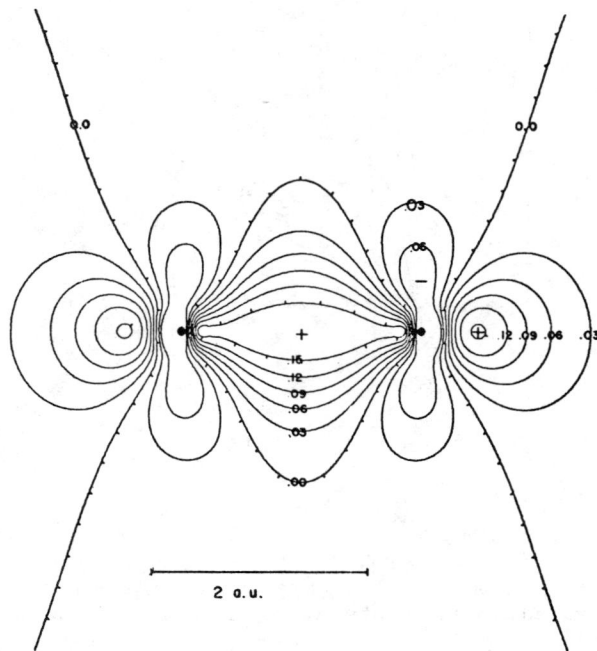

Fig. 10. Density difference map as defined in Fig. Caption 9 with the experimental result replaced by the prediction of HF theory.

ELECTRON COMPTON PROFILE STUDIES

In Figs. 2 and 3 the aforementioned Bethe ridge is the most dominant feature of the experimental surface at high momentum transfers. By use of the binary encounter approximation a relationship can be established between the continuum generalized oscillator strength and the well known x-ray Compton profile, $J(q)$, as[1,17]

$$J(q,K) = \frac{2K^3}{E} \frac{df(K,E)}{dE} \quad (19)$$

and

$$\lim_{K \to \infty} J(q,K) \to J(q) = 2\pi \int_q^\infty dp\, p\, \rho(p) \quad (20)$$

where $q \doteq (E-K^2)/2K$ and $\rho_0(p)$ is the spherically averaged three dimensional one electron momentum density. Eq. 19 makes it possible to define an effective Compton profile even when the limit is not approached. However in order to possess some physical significance the result should be nearly independent of K over some finite range of K.

The use of fast electrons to measure Compton profiles can be traced back nearly 40 years. However those brief efforts appear to have been overlooked until new experiments were performed in 1973.[18] Modern experimental techniques have made it possible to obtain results at nearly the 1% accuracy level. New experiments to be reported at this meeting by Prof. Wellenstein's Brandeis group go well beyond that level.

It is of interest to explore the meaning of Eq. 19 from an experimental point of view. In Figs. 11 and 12 $J(q,K)$ for N_2 and Ar obtained with 25 keV electrons are plotted.[19] The vertical scale is absolute by virtue of the use of Bethe sum rule normalization. The behavior of $J(q,K)$ shown in Figs. 11 and 12 provides a striking example of the importance of shell structure in both atoms and molecules. Note that within the several percent experimental error it is possible to define cumulative shell Compton profiles. That is in the case of N_2 for scattering angles less than about 5° (at 25 keV) it appears possible to define a valence shell Compton profile since the region near the center of the Bethe ridge is devoid of inner shell contributions. After the core shell spectrum in N_2 which starts at an energy loss of 400 eV, passes through the center of the Bethe ridge (actually the ridge passes through the inner shell ionization region) a new profile is produced which now includes the core contribution and agrees well numerically with theoretical values computed from Eq. 20. In the case of Ar the same behavior is observed except that the lower profile is produced by electrons in the M shell and the upper by electrons in the M + L shells. The K shell contribution begins at an energy loss of 3,200 eV and is still quite well separated even at 12° where the peak center still occurs at an energy loss of less than 1000 eV. Detailed calculations of the valence profiles using Eq. 20 but putting in the momentum density only for the electrons belonging to the shell in question gives agreement at about the 4% level with experiment. Better agreement (1-2%) is obtained by calculations based on Eq. 19.[7]

In Tables III and IV recently determined values for all the finite integer power moments of the momentum distribution (all 7) are given for He and H_2.[20] Also included are values for the momentum density and its curvature at zero momentum. The agreement between theory and experiment is excellent. The current experiments with the new equipment in Prof. Wellenstein's laboratory at Brandeis University promises to yield even more accurate results.

The Compton profile measurements as pointed out above yield information on the momentum distribution in a rigorous manner only in the limit of large momentum transfer which is in the language of this conference the "knockout reaction" limit. It should be possible to compare (e,2e) experimental information with Compton profile results in this limit. So far this has only been done for H_2 but the agreement was extremely good. The absolute accuracy of Compton measurements appears to be 3 times better than current (e,2e) results but that advantage is completely lost when the Compton results are transferred to a momentum density so that the shapes of the total momentum density are given about equally well by the two methods. Of course the (e,2e) results give far more detailed information since orbital momentum densities are measured directly.

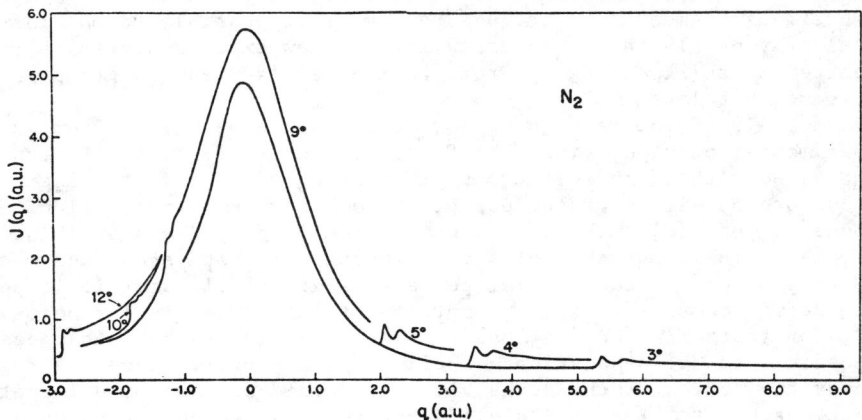

Fig. 11. A plot of $J(q,K)$ as a function of q for various scattering angles as determined by 25 keV electrons scattered from N_2. Bethe sum rule normalization was used to place the values on an absolute scale. Note that the momentum transfer is proportional to the scattering angle.

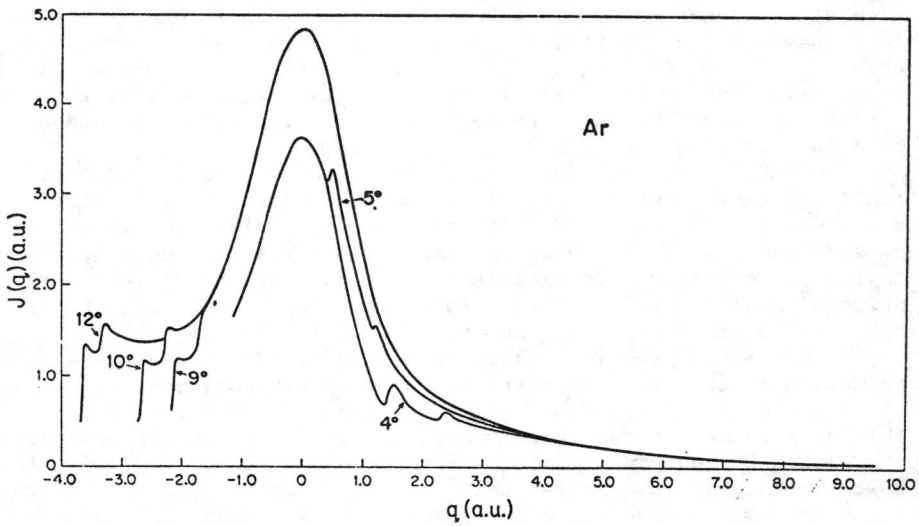

Fig. 12. A plot of $J(q,K)$ as a function of q for various scattering angles as determined by 25 keV electrons scattered from Ar. Bethe sum rule normalization was used to place the values on an absolute scale.

Table III. Results for the Compton Profile of He[1] at ~ 10°

Trial	1	2	3	4	Average	Correlated	RHF	He[4]
$\langle p^{-2} \rangle$	4.090	4.114	4.138	4.119	4.115 ± 0.144	4.114	4.095	
$\langle p^{-1} \rangle$	2.128	2.129	2.128	2.128	2.128 ± 0.016	2.137[3]	2.141[3]	2.141
$\langle p^{0} \rangle$	1.998	2.001	1.990	1.990	1.995 ± 0.020	1.997[3]	2.000[3]	
$\langle p^{1} \rangle$	2.836	2.830	2.825	2.821	2.828 ± 0.042	2.818	2.798	2.799
$\langle p^{2} \rangle$	5.867	5.757	5.658	5.809	5.818 ± 0.291	5.799[3]	5.709[3]	5.723
$\langle p^{3} \rangle$	17.63	16.59	17.22	17.01	17.11 ± 3.42	17.64	17.78	
$\langle p^{4} \rangle$	75.79	64.47	64.61	63.35	67.05 ± 67.0	82.0	100.9	105.86
$\rho(0)$	0.447	0.462	0.463	0.455	0.457 ± 0.078	0.448	0.440	
$\rho''(0)$	-1.74	-2.18	-1.975	-1.821	-1.93 ± 1.93	-1.713	-1.62	
$\sigma^2 \times 10^6$	4.328	6.785	5.994	3.228		1.0068×10^{-2}	4.436×10^{-5}	

[1] Moments were obtained by fitting the Compton profile with 5 terms of the form $\sum_{n=1}^{5} a_n/[1 + (q/\xi_n)^2]^{n+2}$

[2] Moments were obtained by fitting the results of R. Benesch (submitted for publication in J. Phys. B.:Atom. Mol. Physics).

[3] The $\langle p^0 \rangle$, $\langle p^2 \rangle$, $J(0) = \frac{1}{2}\langle p^{-1} \rangle$, given by the authors are 1.997, 5.803, 1.068$_5$ with the function given above in (1) for the correlated case and 2.000, 5.680, and 1.0705, for the RHF case (function of Clementi).

[4] Moments calculated by Sahni, et al., (Phys. Rev. A **12**, 768 (1975)) using a six-parameter analytic Hartree-Fock wave function of P. S. Bagus and T. L. Gilbert.

Table IV. Results for the Compton Profile of H_2[1] at $\sim 10°$

Trial	1	2	3	4	5	Average	Theoretical[2]
$\langle p^{-2} \rangle$	8.312	8.308	8.326	8.348	8.302	8.319 ± 0.249	8.022
$\langle p^{-1} \rangle$	3.075	3.080	3.085	3.077	3.078	3.079 ± 0.021	3.058[3,4]
$\langle p^{0} \rangle$	1.995	1.997	1.995	1.986	1.990	1.993 ± 0.020	2.000
$\langle p^{1} \rangle$	1.864	1.861	1.859	1.835	1.850	1.854 ± 0.028	1.857
$\langle p^{2} \rangle$	2.370	2.358	2.378	2.277	2.345	2.346 ± 0.094	2.342[3]
$\langle p^{3} \rangle$	4.092	4.067	4.244	3.751	4.088	4.048 ± 0.405	4.126
$\langle p^{4} \rangle$	10.51	10.55	12.12	8.68	11.06	10.58 ± 5.29	12.02
$\rho(0)$	1.257	1.250	1.246	1.272	1.244	1.254 ± 0.251	1.149
$\rho''(0)$	-9.20	-8.99	-8.69	-9.69	-8.77	-9.07 ± 9.1	-7.058
$\sigma^2 \times 10^{+5}$	2.578	4.356	5.645	3.020	1.659		0.0132

[1] Moments were obtained by fitting the Compton profile with a 2 term function of the form $\sum_{n=1}^{2} a_n/[1 + (q/\xi_n)^2]^{n+2}$

The average x-ray incoherence scattering factor from this five set of data is $S(K) = 1.996$.

[2] Moments were obtained by fitting the results of R. Brown and V. Smith, Phys. Rev. A 5, 5 (1972). Note that their values do not include vibrational corrections.

[3] The values of $J(0) = \frac{1}{2}\langle p^{-1}\rangle$ and $\langle p^2 \rangle$ given by the authors are 1.529 and 2.348 respectively.

[4] A calculated value for $\langle J(0) \rangle$ using the Bown Liu wave function for H_2 with a vibrational correction yields the result $\langle J(0) \rangle_0 = 1.5467$. Private communication, Prof. V. H. Smith, Jr.

SPECTROSCOPIC INVESTIGATIONS

The major advantage of the (e,e′) method as a spectroscopic tool over photoabsorption or Raman scattering techniques is that it is a relatively simple matter to obtain the momentum transfer dependence of individual energy loss peaks. The dependence often contains important clues as to the origin of the transition under investigation. A number of examples recently obtained in the Indiana University Electron Impact Laboratory follow.

If one observed the angular dependence of the lowest lying optically allowed transitions in the rare gases Ne through Xe ($np^5(n+1)s \leftarrow np^6$) with 25 keV electrons the results shown in Fig. 13

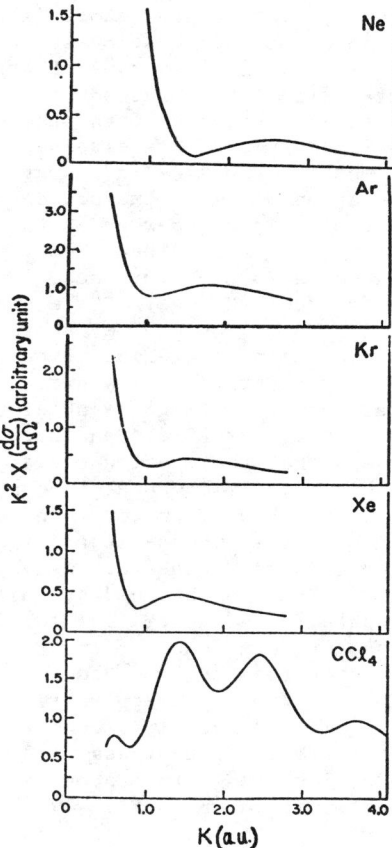

Fig. 13. The momentum transfer dependence of the $np^5(n+1)s \leftarrow np^6$ transitions in the rare gases Ne through Xe as obtained by 25 keV electron scattering.

are observed.[21] The lowest curve labeled CCl_4 is the elastic scattering from carbon tetrachloride which provides a calibration of the momentum transfer scale since the positions of the maxima are well understood. Note that all the rare gases shown display a definite minimum and maximum. This behavior is the result of interference between the nodal structure of the radial transition charge density and the spherical Bessel function of order 1 which occur in the Born radial matrix element.[22] In Table V the observed minima and maxima positions are compared with previous experimental results for Kr and Xe and with theoretical calculations based on relativistic Hartree-Fock wave functions for the atom. The pleasing agreement argues in favor of using such measurements as a diagnostic tool for making spectral assignments.

If the focus is shifted to the onset of inner shell excitation processes in molecules containing complex atoms (i.e. more than one shell) as for instance the K shell ionization structure observable at 400 eV in N_2 as shown in Fig. 3 a new type of structure may be observed. In this case it suffices to focus on the small angle spectra obtained in the K and L shell regions of S and the K shell region of F in the molecule SF_6. These spectra are displayed in Figs. 14-16.[23] The energy scale corresponds to the kinetic energy of an ejected electron from the edge in question assuming the process is a simple single ionization. The actual energy loss observed in the experiment is obtained by adding the ejected electron kinetic energy to the ionization potential at the edge which is 2490 eV for the S-K edge, 180 eV for the S-L edge and 695 eV for the F-K edge. These spectra show a number of shape resonances in the region 0-50 eV above the edge[24] and several weak maxima and minima in the region beyond which can be attributed to extended x-ray absorption fine structure (EXAFS).[25] Figs. 14-16 are the first such results to be obtained from electron scattering by a gaseous target. EXAFS are another example of shape resonances. It is supposed that the ejected electron in the ionizing collision will be elastically backscattered by a neighboring atom which will form resonance states whenever an integer multiple of the electron wavelength matches the nearest neighbor distance. This means that useful information on nearest neighbor distances to a specific elemental type can be obtained by suitable Fourier transformation of EXAFS data.[25] A future goal for investigation will be the momentum transfer dependence of the EXAFS structure. This information is unique to the HEEIS method.

A final example of a spectroscopic application is provided by observation of the continuum near the first ionization threshold as a function of momentum transfer. In the case of 25 keV electrons scattered by N_2 the energy loss range from about 8 eV to 50 eV over a range of momentum transfer is shown in Fig. 17.[26] The 1° scattering angle is sufficiently small so that the intensity closely approximates that obtained in a photoabsorption experiment with an energy resolution of about 2 eV. As the maximum in the ionization continuum moves to higher energy loss (the Bethe ridge) the region left behind displays a new transition(s) not observed in the photoabsorption cross section. This broad transition has been assigned as a

TABLE V. Experimental values of the momentum transfer K (a.u.) for minima and maxima in $f(K,E_n)$ in the rare gas $[np^5)(n+1)s]$ transitions, and comparison with previous experimental and theoretical results.

Atom	Minimum		Maximum	
	Exper.	Theory	Exper.	Theory
Ne [2p)3s]	1.63 ± 0.05	1.7[b]	2.72 ± 0.10	2.6[b]
		2.04[c]		3.39[c]
Ar [3p)4s]	1.12 ± 0.05	1.2[b]	1.72 ± 0.10	1.8[b]
		2.7[c]		3.01[c]
		1.20[d]		1.69[d]
Kr [4p)5s]	1.02 ± 0.05	1.1[b]	1.58 ± 0.10	1.6[b]
	0.95[a]	0.83[c]		1.16[c]
		1.10[e]		
Xe [5p)6s]				
$^1P_1$		0.97[b]		1.4[b]
		0.99[c]		1.16[c]
	0.94 ± 0.05	0.93[e]	1.38 ± 0.10	1.4[e]
	0.84[a]		1.20[a]	
$^3P_1$		0.98[b]		1.4[b]

[a] Y.-Kim, M. Inokuti, G. G. Chamberlain and S. R. Mielczarek, Phys. Rev. Lett. 21, 1146 (1968).

[b] Y.-K. Kim, private communication. Calculated from relativistic Hartree-Fock wavefunctions.

[c] K. J. Miller, J. Chem. Phys. 59, 5639 (1973).

[d] Values obtained from Fig. 3 of R. A. Bonham, J. Chem. Phys. 36, 3260 (1962), where momentum transfer, K, was given in units of $Å^{-1}$. These values are based on calculations made with numerical Hartree-Fock wavefunctions and were misquoted in Table III of K. J. Miller, J. Chem. Phys. 59, 5639 (1973) as being a.u..

[e] Theoretical values quoted in Y.-Kim, M. Inokuti, G. G. Chamberlain and S. R. Mielczarek, Phys. Rev. Lett. 21, 1146 (1968), based on calculations made with Hartree-Fock wavefunctions.

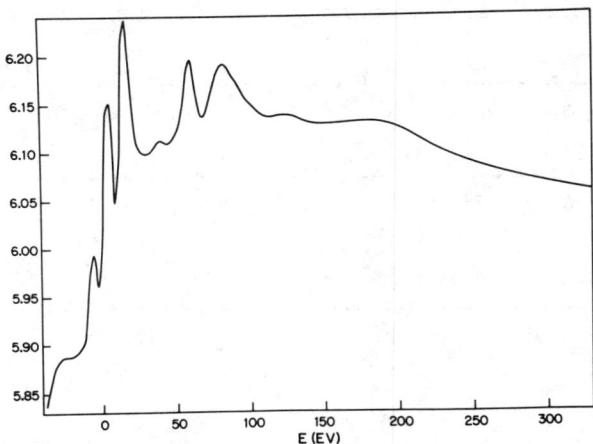

Fig. 14. Plot of the K shell ionization region for S in SF_6. The vertical scale is relative, the incident energy was 25 keV and the scattering angle was 1°. The intensity was leveled by multiplication by the energy loss. The horizontal scale is the difference between the actual energy loss and the K shell ionization threshold at 2490 eV.

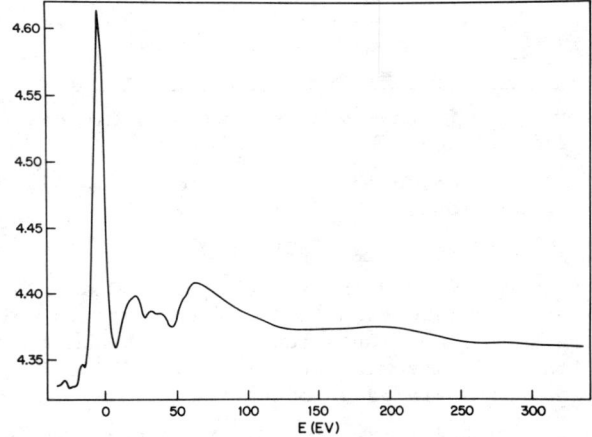

Fig. 15. Plot of the L shell ionization region for S in SF_6. The vertical scale is relative, the incident energy was 25 keV and the scattering angle was 1°. The intensity was leveled by multiplication by the energy loss raised to the 3.2 power. The horizontal scale is the difference between the actual energy loss and the $^2P_{3/2}$ ionization edge at 180 eV.

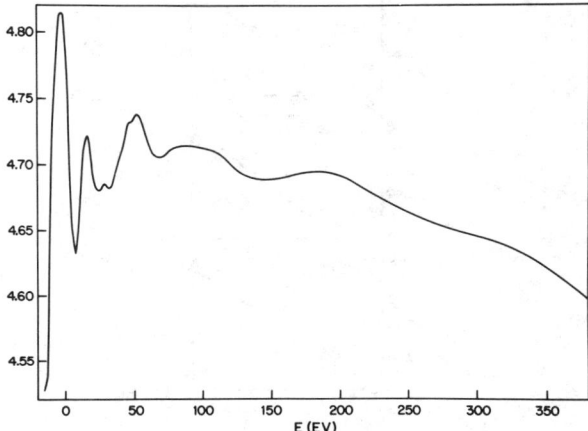

Fig. 16. Plot of the K shell ionization region for F in SF_6. The vertical scale is relative, the incident energy was 25 keV and the scattering angle was 1°. The intensity scale was leveled by multiplication by the energy loss raised to the 5.5 power. The horizontal scale is the difference between the actual energy loss and the F-K ionization edge at 694.6 eV.

transition from an inner molecular orbital (MO) to the lowest unfilled MO. This corresponds to the transition $^1\pi_g(2s\sigma_g^{-1}2p\pi_g^{1}) \leftarrow$ $^1\pi_g$ which is a dipole forbidden but quadrupole allowed state. The unusual angular dependence of this state is characteristic of that expected for a quadrupole transition. Recently Dr. Lee[27] in my laboratory has observed additional transitions of this type in the molecules O_2, NO, CO, CO_2 and N_2O. In Fig. 18 the angular or momentum transfer dependence of the NO spectrum is given. Note that at least two new peaks are observed in the large angle spectrum. These results are not yet completely understood but clearly demonstrate the utility of the (e,e′) method in obtaining new information on the electronic structure of the target.

74

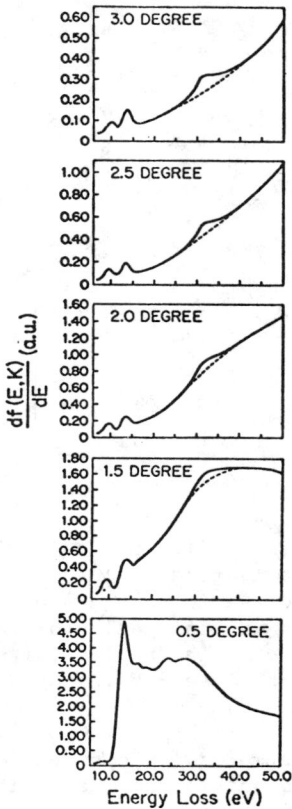

Fig. 17. The angular dependence of the energy loss spectrum for N_2 obtained by 25 keV electron scattering over the energy loss range of 8 eV to 50 eV with a resolution of 2 eV. The vertical scale is absolute and was determined by use of the Bethe sum rule for the entire energy loss spectrum. The new peak in the continuum appears to be centered at an energy loss of 31 eV.

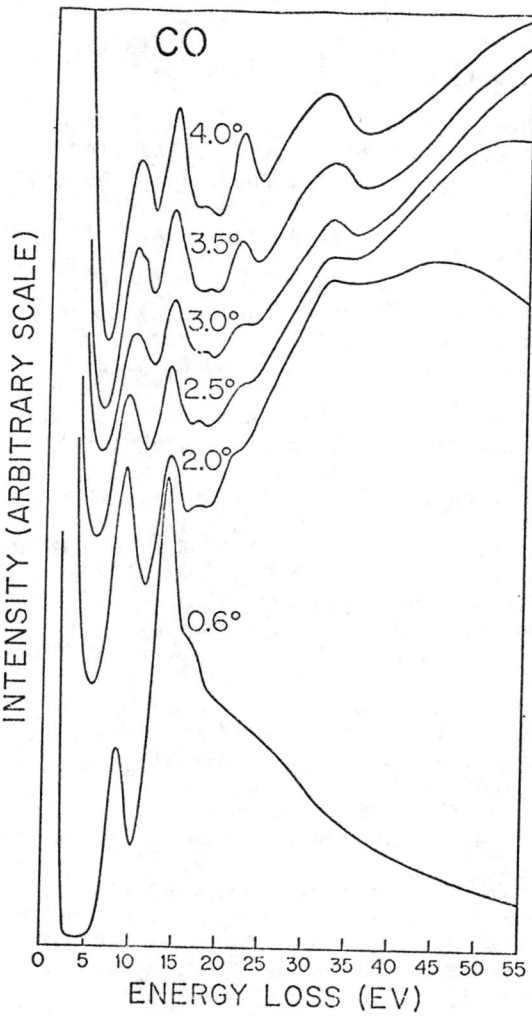

Fig. 18. The angular dependence of the energy loss spectrum for CO obtained by 25 keV electron scattering over the energy loss range 8 eV to 50 eV with a resolution of 2.5 eV. The vertical scale is relative and the two new peaks in the continuum are centered at 21.7 eV and 31 eV.

REFERENCES

1. M. Inokuti, Rev. Mod. Phys. 43, 297 (1971).
2. H. Schmoranzer, H. F. Wellenstein and R. A. Bonham, Rev. Sci. Instr. 46, 89 (1975).
3. H. F. Wellenstein, H. Schmoranzer, R. A. Bonham, T. C. Wong, and J. S. Lee, Rev. Sci. Instr. 46, 92 (1975).
4. H. F. Wellenstein, J. Appl. Phys. 44, 3669 (1973).
5. U. Fano and J. W. Cooper, Rev. Mod. Phys. 40, 441 (1968).
6. E. N. Lassettre, A. Skerbele and M. A. Dillon, J. Chem. Phys. 50, 1829 (1969).
7. B. J. Block and L. B. Mendelsohn, Phys. Rev. A9, 129 (1974).
8. R. A. Bonham and M. Fink, "High Energy Electron Scattering", (Van Nostrand-Reinhold, 1974, New York).
9. R. C. Ulsh, H. F. Wellenstein and R. A. Bonham, J. Chem. Phys. 60, 103 (1974).
10. H. F. Wellenstein, H. Schmoranzer, R. A. Bonham, T. C. Wong and J. S. Lee, Rev. Sci. Instr. 46, 92 (1975).
11. J. S. Lee, unpublished results.
12. S. M. Silverman and E. N. Lassettre, J. Chem. Phys. 42, 3420 (1965).
13. L. C. Lee, R. W. Carlson, D. L. Judge, and M. Ogawa, J. Quant. Spectrosc. Radiat. Transfer 13, 1023 (1973).
14. C. Tavard, Cah. Phys. 20, 397 (1965).
15. M. Fink and R. A. Bonham, Phys. Rev. 187, 114 (1969).
16. M. Fink, D. Gregory and P. G. Moore, Phys. Rev. Lettrs. 37, 15 (1976).
17. R. A. Bonham and C. Tavard, J. Chem. Phys. 59, 4691 (1973).
18. H. F. Wellenstein and R. A. Bonham, Phys. Rev. A7, 1568 (1973).
19. T. C. Wong, J. S. Lee, H. F. Wellenstein and R. A. Bonham, Phys. Rev. A12, 1846 (1975).
20. J. S. Lee, J. Chem. Phys. (in press).
21. T. C. Wong, J. S. Lee and R. A. Bonham, Phys. Rev. A11, 1963 (1975).
22. R. A. Bonham, J. Chem. Phys. 36, 3260 (1962).
23. W. St. John, R. E. Kennerly and R. A. Bonham (to be published).
24. J. L. Dehmer, J. Chem. Phys. 56, 4496 (1972).
25. D. E. Sayers, E. A. Stern and F. W. Lytle, Phys. Rev. Lettrs. 27, 1204 (1971).
26. J. S. Lee, T. C. Wong and R. A. Bonham, J. Chem. Phys. 63, 1643 (1975).
27. J. S. Lee, J. Chem. Phys. (in press).

THE IU-MELBOURNE (p,2p) EXPERIMENTS AT IUCF

D.W. Devins
Indiana University Cyclotron Facility, Bloomington, IN 47401

ABSTRACT

The motivation for, planning of and current status of the (p,2p) experiments at IUCF are discussed.

INTRODUCTION

The utility of the (p,2p) reaction as a spectroscopic probe has been too thoroughly discussed at this meeting to warrant further elaboration here. However, in view of the fact that experiments of this type at energies above about 100 MeV have not been as successful as had been hoped, it is necessary to mention the reasons for attempting them again at IUCF.[1] The best summed energy resolution obtained at 155 MeV[2] and 185 MeV[3] has varied between 1.5 and 3.0 MeV. This means that for all targets studied (except ^{6}Li) it has not been possible to resolve individual energy levels of the residual nucleus formed by proton knock out, even for p-shell targets. Spectroscopic information has been limited to separation energies of the centers-of-gravity of levels from various shells up to about Z=30 and angular correlations of groups of levels.

The reasons for the lower quality data at higher energies stem from the rather low cross section for (p,2p) reactions, typically a few tens of $\mu b/Sr^2 MeV$, coupled with the low intensity, low duty factor and poor quality of beams from typical synchrocyclotrons. These factors necessitate a tradeoff between count rate and resolution, usually in favor of the former at the expense of the latter. Typically large solid angles are used introducing a smearing over recoil momentum and, because of the penetrating power of higher energy protons, relatively low resolution ($\sim$ 1%) plastic or NaI(Tl) detectors were used in detector telescopes.

The IU-Melbourne group plans to capitalize on the unique beam qualities of the IUCF machines and use the latest semiconductor detector technology along with a magnetic spectrograph to obtain (p,2p) summed energy spectra about an order of magnitude better in resolution than those obtained earlier in the neighborhood of 150 MeV.

EXPERIMENTAL DETAILS

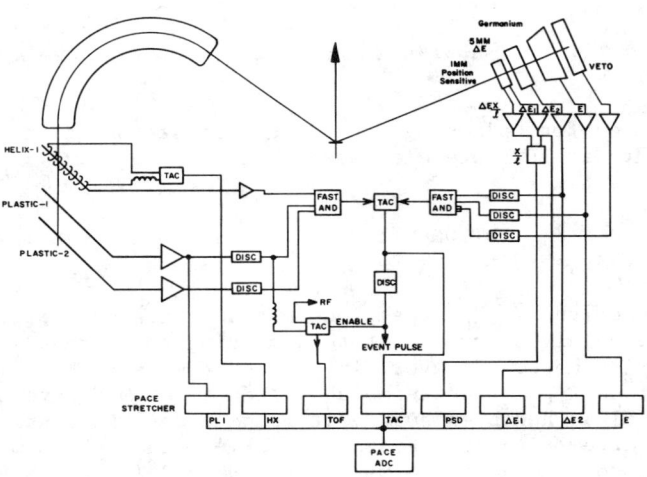

Fig. 1. Simplified block diagram of the detection-electronics system to be used in the knock-out experiments at IUCF.

Figure 1 is a simplified block diagram of the detection-electronics system to be used in the knock-out experiments at IUCF. The spectrograph arm has a solid angle of about 2.1 msr with an angular aperture of 2.4°. Momentum measurement is accomplished with a 50 cm helical wire position sensitive chamber developed by the Melbourne group.[4] The best overall resolution measured with the helical counter has been $\sim$ 50 keV. The intrinsic resolution of the counter is probably much better, but the 50 keV figure includes the effects of the magnet as well. Two plastic scintillators are used to reduce background. One is also used to measure the time of flight of each proton through the spectrograph relative to the RF pulse. The kinematic correcting coils are set for k=0 so that the time of flight uniquely measures the path taken through the magnet and, hence, the scattering angle. The FWHM of the time of flight peak has been measured to be 0.7 ns, which corresponds to an angular error of 0.6°.

The original plans for the detector telescope called for a 1 mm Si-surface barrier transmission mounted position sensitive counter followed by a 5 mm Si-surface barrier ΔE counter and a 1 cm thick 3 cm diameter intrinsic germanium stopping detector, as indicated in Figure 1. Unfortunately, the three germanium detectors which were prepared for us two years ago by the Pehl group at Berkeley are not currently operable. We are trying to have one refurbished, but it is not clear when that operation will be complete. In the meantime, we are preparing to operate at lower incident energies, near 100 MeV, using a stack of 5 mm counters thick enough to stop protons of $\sim$ 44 MeV. The telescope solid angle is 3.5 msr with an angular opening of $4.5°$. The position counter resolution is 0.5 mm corresponding to $0.4°$. The overall energy resolution of the telescope has been measured to be 75 keV FWHM.

Table 1. Resolution

Contributor	Contribution (keV)	
	Best Case[a]	Worst Case[b]
I. Beam		
Quality		80
Spot Size[c]	0	60
Divergence	<5	<5
II. Target[d]		
Straggling		30
Multiple Scattering	20	30
Event location	60	90
III. Detectors		
Finite Solid Angle[c]	0	60
Instrumental (Spectrograph)	30	
(Telescope)	75	

a) $\theta_1 = \theta_2 = 47°$

b) $\theta_1 = \theta_2 = 65°$

c) Assumes spot size = 2 mm, $\Delta\theta = 1/2°$

d) Assumes target = 2.2 mg/cm^2 $\theta_t = 0°$

Table 1 contains a listing of the measured or calculated contributions of various factors to the experimental resolution for the proposed knock-out experiments. If these were the only factors involved, the resolution would vary between 160 and 200 keV. Unfortunately, the quantity of interest in the experiment is the summed energy spectrum. Each j-th event in this spectrum is the sum of four detector outputs (Helix, ΔE_1, ΔE_2 and E), each of which has been converted to energy by application of a linear calibration:

$$E_{sum}^j = \sum_{i=1}^{4} m_i x_{ij} + b_i \quad , \qquad (1)$$

where x_{ij} is the ADC channel number of the i-th detector for the j-th event.

The calibration factors m_i and b_i each have associated errors, since they are obtained experimentally. For a specified knock-out event (i.e., given $\vec{p}_1$, $\vec{p}_2$, $\vec{q}$, event location in the target, etc.), the sum in quadratures of the factors in Table 1 must represent the error associated with the x_{ij}. However it is not obvious that the error in E_{sum}^j can be obtained by summing the errors in m_i, x_{ij}, and b_i appropriately. There is no distribution of values of m_i and b_i for a particular detector; all events in the i-th detector producing an ADC signal in channel x will correspond to the same energy, E_i, in that detector. There are two facts which make error analysis possible. The same E_{sum}^j can be produced with an infinite combination of E_i's and two events producing the same x in the i-th ADC may have rather different x's in the other three ADC's. Since the absolute errors in the m_i and b_i are channel dependent, these calibration errors will produce a distribution in summed energies for a succession of identical knock-out events (I will not try to argue today that the distribution is Poisson). Therefore, I will assume that the various errors should be added appropriately.

The conversion factors are obtained by scattering from hydrogen, calculating the expected energy signal for each detector, then using a Gaussian fitting routine[5] to obtain the peak channels. A calibration curve from which m_i and b_i are deduced for each detector is made from the E_{calc} vs Ch_{peak} points. Experimental errors can be introduced at several places: beam energy, angle setting, detector thickness, energy loss calculations, peak fitting and curve fitting. Preliminary estimations are that these errors would amount to about 0.3% for each detector, or a total of about 250 keV. This single effect is much larger than any of the instrumental effects of Table 1, and will be difficult to reduce with the current arrangement.

It should be noted that the values for the errors associated with beam spot size and finite solid angle assumed an angular resolution of 0.5°. Since the angular openings of both detector systems are larger than this, we must measure the actual scattering angles of each proton in each event using the time of flight and position sensitive detectors previously described. Then we calculate the recoil momentum for each event, using conservation of momentum and energy, and add it to the summed energy for that event. This recoil effect has so far been small, since it has been masked by the larger calibration error.

Even at the currently calculated resolution, 320 to 360 keV FWHM, the summed energy spectra will be from 2 1/2 to 3 times better resolved than the best work reported to date at 100 MeV.[6] At 150 MeV we should have from 5 to 10 times better resolution than heretofore available.

The detector scheme shown in Figure 1 has several advantages and disadvantages over a two telescope arrangement, which has been used in most of the earlier work. The IUCF spectrograph has a rather small energy bite in the focal plane, $\pm$ 3%. At 150 MeV, about 10 magnet settings would be required to cover one half of the E_1 vs E_2 plane in order to map out the recoil momentum distribution as suggested by Phil Roos in his talk. To do this at several angles as well would require an excessive amount of beam time.

A two-telescope detection system would overcome this dynamic energy range problem, but would have other problems. Beam currents would have to be limited to about 100 na at most angles to reduce high count rate induced dead time and resolution degradation effects. Nevertheless, this system would probably be more efficient overall in terms of beam time utilization. Unfortunately, high purity germanium detectors of the depth and area required for this experiment are not readily available, are quite expensive, and, as we have discovered, are not overly reliable. Also the energy lost by protons in excess of 50 MeV in the 1 mm position sensitive counter is small, near the noise threshold of the ORTEC position analyzer module. It may be necessary to devise another technique for measuring position in the telescope. We will attempt in the next year or two to obtain the detectors necessary to mount a two-telescope arrangement.

INITIAL MEASUREMENTS

Proton beams from $\sim$ 100 to $\sim$ 160 MeV constitute an important part of the current physics program at IUCF. Before these beams can be used effectively in nuclear structure studies, the important aspects of the reaction mechanisms effective at these intermediate energies must be elucidated. For many years the DWBA using optical model generated scattered waves has been used in the analysis of reaction data. This procedure has met with some success at lower energies in providing spectroscopic information, although the reliability of this information is uncertain. Particularly with composite particles, but even with "elementary" probes, spectroscopic factors for the same final states may vary widely as a function of probe and incident energy. Many factors may contribute to the difficulties with the DWBA. There has always been a question of the importance of multiple-step processes in nuclear reactions, and more recently the existence of two-step mechanisms has been determined.[7]

The extent to which two-step processes are important at IUCF energies is essential to establish: if they are important, they will complicate reaction formulations and make reliable spectroscopic information extraction difficult; at the very least, their role must be quantitatively understood or spectroscopic studies at IUCF can yield only qualitative information. The (p,2p) reaction can provide a unique test of two-step reaction mechanism formulations.

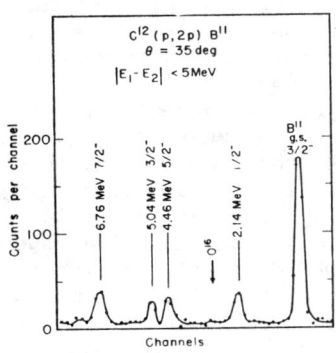

Figure 2. Summed energy spectrum from $^{12}C(p,2p)^{11}B$ at 50 MeV. Reproduced from reference 8.

The reaction $^{12}C(p,2p)^{11}B$ has been studied at several bombarding energies. Only at 50 MeV[8] has the summed energy resolution been good enough to separate the $5/2^-$ state at 4.445 MeV from the $3/2^-$ state at 5.019 MeV. This $5/2^-$ state is unlikely to be formed in a single step direct knock-out and is therefore a good candidate for a two-step process. It is seen with a strength equal to the $3/2^-$ state at 50 MeV, as is apparent from Figure 2. The authors of this work concluded that no reasonable ground state wavefunction for ^{12}C could account for one-step knock-out to the $5/2^-$ and $7/2^-$ states and that two-step processes could be required. However, they did not make any quantitative arguments. Geramb and Strobel[9] have combined one-and two-step mechanisms in a calculation of the $^{16}O(p,2p)^{15}N$ g.s. reaction with some success. They assumed two giant multipole resonances with parameters adjusted to give a good fit. This formulation has not been applied to the two-step states seen in the ^{12}C reaction.

It seems very clear that the nature of two-step interactions at intermediate energies needs to be clarified. The first question we should like to answer is to what extent the ^{11}B two-step states are excited at 100 MeV. If they have a reasonable strength relative to the 3/2⁻, then we can ask the more general question of the energy dependence of two-step reactions. If they are weakly excited, we should perhaps be more concerned with the nature of the intermediate states(s) and their modes of excitation.

Roos has presented some intriguing calculations which we should very much like to follow up on as soon as we have the capability to operate at higher energies. If Roos' DWIA calculations reproduce the data well for $E \geq 150$ MeV, then a large number of nuclear structure problems become amenable to study. Just as interesting is the energy range between 80 and 150 MeV where it is clear the DWIA is not appropriate. In this energy regime (p,2p) data will provide stringent tests for realistic distorted wave reaction mechanism formulations of increasing sophistication. We can look forward to better calculations using fewer approximations of wider validity and utility.

REFERENCES

1. Other members of the IU-Melbourne collaboration are D.L. Friesel, W.P. Jones and A. Attard, I.U.; B.M. Spicer, V.C. Officer, G.G. Shute, I.D. Svalbe and R.S. Henderson, Melbourne University. This collaboration has been supported by the U.S.-Australian Scientific Cooperation Program administered by the National Science Foundation and the Australian Research Grants Committee.
2. For example, Arditi et al., Nucl. Phys. A103, 319 (1967).
3. H. Tyren et al., Nucl. Phys. 79, 321 (1966).
4. V.C. Officer, R.S. Henderson and I.D. Svalbe, Bull. Am. Phys. Soc. 20, 1169 (1975).
5. T. Mukoyama, Nucl. Instr. and Methods, 125, 289 (1975).
6. Bhowmik et al., Phys. Rev. C6, 2105 (1976).
7. For example, Yagi et al., Phys. Rev. Letters 34, 96 (1975).
8. Pugh, et al., Phys. Rev. 155, 1054 (1967).
9. H.V. Geramb and G.L. Strobel, Phys. Letters 34B, 611 (1972).

THE DETERMINATION OF ELECTRONIC MOMENTUM DISTRIBUTIONS
AND ATOMIC AND MOLECULAR STRUCTURE USING THE (e,2e) REACTION

Erich Weigold

The Flinders University of South Australia, Bedford Park, S. A.

ABSTRACT

A survey of symmetric (e,2e) experimental results in atoms and molecules is presented. Information which can be extracted from such experiments using the off-shell impulse approximation is discussed. The noncoplanar symmetric geometry is particularly suited for obtaining structure information, whereas the coplanar arrangement is suited for studying distortion effects. Energy independent structure information is obtained on: (a) shapes and magnitudes of electron momentum distributions in individual electron orbitals; (b) separation energies for ion states with their orbital signature; (c) spectroscopic factors describing the probability that an eigenstate contains the principal configuration of a hole in the characteristic orbital; and (d) details of electron correlation effects in target states as well as final ionic states.

INTRODUCTION

In analogy with the (p,2p) knockout reaction in nuclear physics, the term (e,2e) reaction describes those electron impact ionizing experiments in which the kinematics of the initial and two outgoing electrons are fully determined. Depending on the particular experimental conditions, a large number of different types of experiments can be carried out using the (e,2e) reaction. For instance, Weigold et al [1] used the (e,2e) reaction to study autoionizing states in helium. Since the kinematics were fully determined they were able to gain information on the resonance and direct cross sections and their interference as a function of the momentum $\underline{k}_B$ of the emitted (cascade) electron for known values of the momentum transfer $\underline{K} = \underline{k}_0 - \underline{k}_A$, where $\underline{k}_0$ and $\underline{k}_A$ are the momenta of the incident and scattered (higher energy secondary) electron respectively [Fig. 1].

On the other hand, as Inokuti showed yesterday (see also [2]), at high incident energies and very small momentum transfers K (i.e. $\theta_A \approx 0$) the electron scattering cross section is proportional to the optical oscillator strength for the transition, and the (e,2e) reaction can be used to simulate photoionization. This asymmetric high energy forward scattering arrangement has been used mainly by M. Van der Wiel and C. Brion[3] and co-workers to obtain binding energy spectra and partial oscillator strengths. Some of the results of this work were reported by Inokuti.

Ehrhardt and co-workers[4] have used a low energy asymmetric arrangement in order to test various ionization calculations. Since

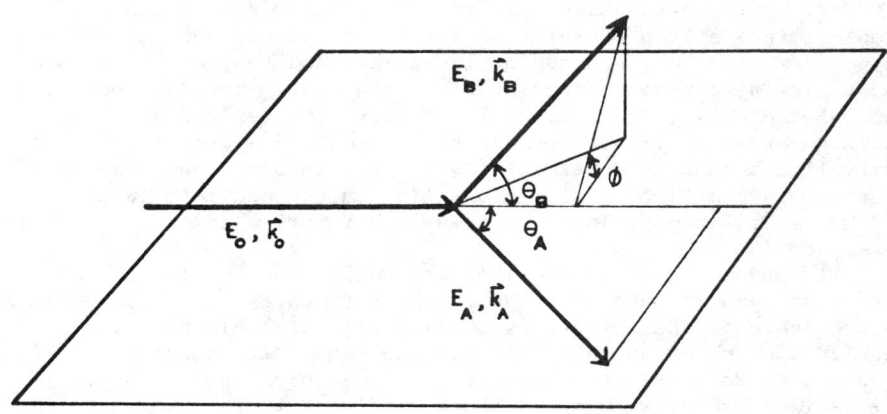

Fig. 1. Scattering kinematics in the (e,2e) reaction. The incident and outgoing momenta are k_O, k_A and k_B respecitvely and $\phi_A = \pi$, $\phi = \phi_B$ measures the deviation from the coplanar geometry.

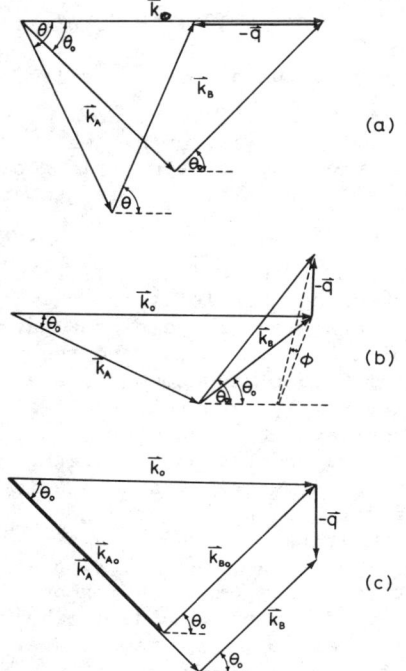

Fig. 2. Possible symmetric kinematical arrangements. θ_0 is the angle for which $q = k_O - k_A - k_B = 0$ when $E_A = E_B$. (a) coplanar symmetric geometry; (b) noncoplanar symmetric geometry and (c) energy varying method.

the asymmetric arrangement emphasizes the long range nature of the
Coulomb interaction between the electrons, binary encounter impulse
type approximations are obviously suspect and they do not at present give a very good description of the data. In addition, since
one of the outgoing electrons is of very low energy distortion
effects are very important, and such experiments are therefore of
very limited use in determining structure information. Obviously
their main aim is to yield information on the collision process.
J. Moore will report later on some recent work carried out in this
area at JILA.

In order to ensure close electron-electron collisions and a
knockout type of mechanism the momentum transfer K must be maximised.
Since the two final electrons are indistinguishable this can be
achieved by choosing the symmetric experimental arrangement in which
the two outgoing electrons are detected at equal polar angles
($\theta_A = \theta_B = \theta$) and with equal energies $E_A = E_B$. This is in contrast
to the distant collisions observed in the case of forward scattering
($K \approx 0$).

In the symmetric arrangement, the magnitude of the ion recoil
momentum $q = \underline{k}_0 - \underline{k}_A - \underline{k}_B$ is given by

$$q = [(2k_A \cos\theta - k_0)^2 + 4k_A^2 \sin^2\theta \sin^2\tfrac{\phi}{2}]^{\tfrac{1}{2}} \qquad (1)$$

where $\phi_A = \pi$ and $\phi_B = \phi$ is the azimuthal variable, the incident beam
defining the Z direction.

There are two symmetric arrangements which are particularly
useful. The first is the coplanar symmetric geometry in which $\phi = 0$
and the angle $\theta = \theta_A = \theta_B$ is varied. This geometry which selects q
parallel to the incident direction k_0 [fig. 2], was the one first
employed by Camilloni, Giardini-Guidoni, Tirribelli and Stefani in
their pioneering work using thin foils.[5] This geometry has also
been used by the Flinders group who employed gas targets to study
the valence states of the inert gases[6] and recently also with gas
targets by A. Giardini-Guidoni and co-workers at Frascati[7]. This
geometry has been extensively used to investigate (p,2p) reactions.

In the noncoplanar symmetric geometry, on the other hand,
$\theta_A = \theta_B = \theta$ is kept fixed and the azimuthal angle ϕ is varied. This
geometry, which selects values of q which lie in the plane perpendicular to $\underline{k}_0$, was first used by Weigold, Hood and Teubner.[8] This
experiment was also the first to study individual valence orbitals
and demonstrate the importance of correlation effects in the final
state. There is a third arrangement, which one can call the energy
varying or sharing method, which should also be very useful. In
this arrangement $\phi = 0°$ and $\theta_A = \theta_B$ are fixed at θ_0 (the angle for
which q = 0 when $E_A = E_B$) and E_A and E_B are varied keeping their sum
E constant (see fig. 2). This arrangement, which has yet to be used
in the (e,2e) reaction, has been used successfully to study (p,2p)
processes. As pointed out by I. E. McCarthy in his opening talk,

at high enough energies the cross section in noncoplanar symmetric geometry depends only on the Fourier transform of the overlap between the target and final ion state, the half-off-shell Mott scattering cross section being nearly independent of the angle ϕ. This, and the fact that the experimental distribution must be symmetric about the planar position ($\phi = 0$) makes this geometry particularly useful for measuring momentum distributions and determining structure information. I will concentrate here on the noncoplanar experiments, which account for the bulk of the momentum distribution and structure work carried out to date. Some of the Flinders coplanar symmetric data have already been discussed by I. E. McCarthy with particular reference to distortion effects at angles $\theta < 45°$. I will discuss this further on Thursday morning. An additional reason for omitting the coplanar work is that A. Giardini-Guidoni will shortly present the Frascati results.

EXPERIMENTAL TECHNIQUES

A schematic diagram of the noncoplanar and coplanar apparati currently being used at Flinders is shown in figures 3 and 4 respectively. The noncoplanar symmetric spectrometer consists of two cylindrical mirror analyzers mounted so that the entrance angle relative to the incident direction (42.3° in this case) is approximately θ_0 (which is always $\leq 45°$). Alignment of the incident beam from an electron gun with the interaction region is achieved by means of two orthogonal pairs of deflection plates. The gun and deflection plates are mounted in a stainless steel tube in order to shield the interaction region from stray electrostatic fields and secondary electrons. Focussing of the beam is achieved with the aid of the Faraday cup by maximising the ratio of the current through the cup aperture to the current collected on the aperture plate.

High purity gas samples are passed through a variable leak valve and then through a stainless steel nozzle into the interaction region. Electrons from the interaction region are decellerated by cylindrical retarding lenses and focussed on the entrance aperture of each cylindrical mirror analyzer. The acceptance angle of each analyzer is approximately 4°. Behind the exit aperture of each analyzer is mounted a grid for suppressing secondary electrons and a channel electron multiplier. One of the analyzers can be rotated about the beam direction by means of a stepping motor mounted outside the vacuum system. As usual care must be taken to shield against magnetic and stray electric fields.

A schematic diagram of the standard coincidence and signal processing electronics used is shown in fig. 5. Pulses from the channel electron multipliers (CEM) pass through preamplifiers, double delay line amplifiers (DDL) and cross over pick-off fast timing discriminators. One of the fast discriminators starts the time to amplitude converter (TAC), the stop pulse being suitably delayed. The TAC output is fed simultaneously to two single channel analyzers and a multichannel analyzer. The window of one SCA is set over the coincidence peak, the second SCA recording accidental events. The

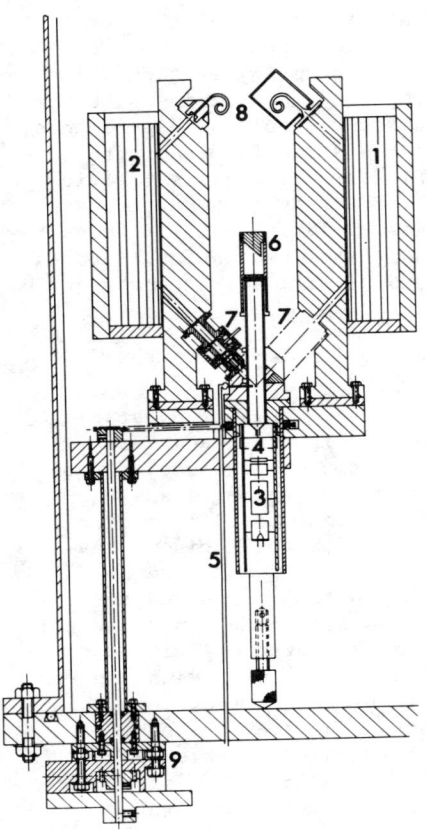

Fig. 3. The noncoplanar coincidence spectrometer. 1 & 2 - rotatable (fixed) cylindrical mirror analyzers; 3 - electron gun; 4 - deflection plates; 5 - gas inlet tube; 6 - Faraday cup; 7 - retarding lenses; 8 - channel electron multipliers and grids; 9 - rotatable feed through.

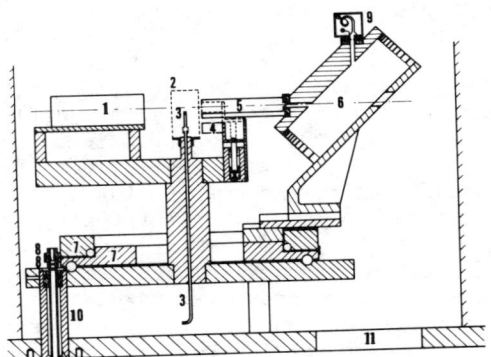

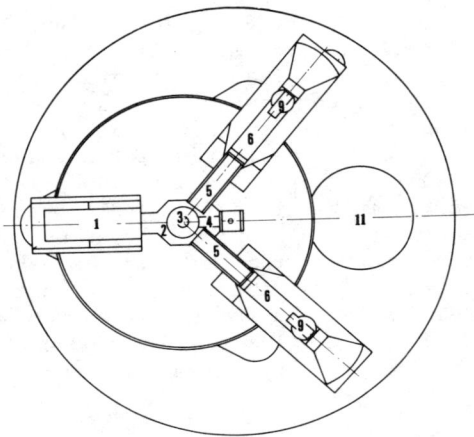

Fig. 4. Side and top elevations of the coplanar coincidence spectrometer. 1 - electron gun and mount; 2 - shielding ring; 3 - gas inlet; 4 - retractable Faraday cup; 5 - collimators; 6 - quadrupole deflectors and retarding lenses; cylindrical mirror analyzers; 7 - turntables; 8 - gears; 9 - channel electron multipliers; 10 - rotating feedthrough; 11 - pumping port.

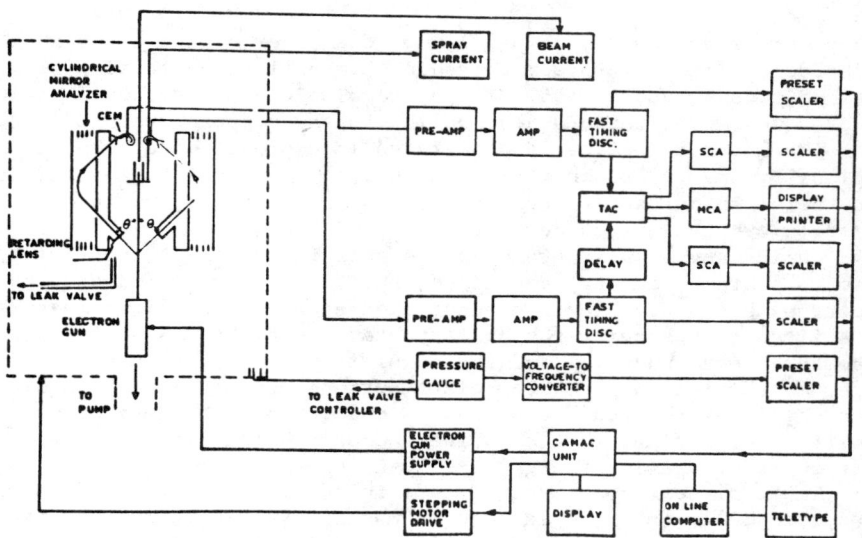

Fig. 5. Schematic diagram of electronic circuitry.

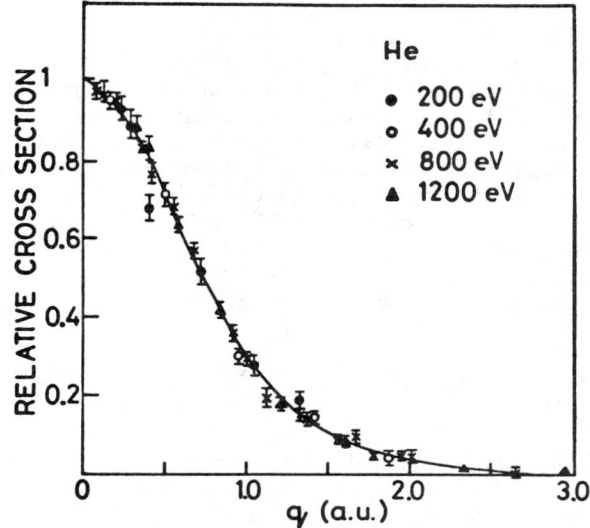

Fig. 6. The noncoplanar differential cross section for the He (e,2e) He$^+$(1s) reaction plotted as a function of q. The solid line is the Hartree-Fock $|\phi_{1s}(q)|^2$.

background window width is usually of the order of ten times the coincidence window in order to reduce statistical uncertainties in the background subtraction. Timing resolutions are typically 5-10 nanoseconds. The third scaler online to the computer records counts from the movable analyzer and operates in a preset mode. In the noncoplanar geometry its count rate should be independent of ϕ and it can therefore be used to correct for any small variation in the electron current or gas density.

During any experimental run the energies of the two outgoing electrons are kept equal and fixed, the total energy of the outgoing electrons being E. Electron separation energies are first measured at a fixed ϕ by varying the primary beam energy by means of a programmable power supply. For the angular correlation measurements (i.e. measurement of momentum distributions) the azimuthal angular setting of the rotatable analyzer is changed by the computer controlled stepping motor, the incident energy being fixed to give a particular separation energy $\epsilon_i = E_0 - E_A - E_B = E_0 - E$.

The coplanar experiment differs in several respects. These are discussed fully in [9]. In general much greater care must be taken to ensure that the coincidence efficiency is independent of the angular setting.

THEORETICAL BACKGROUND

The basic (e,2e) approximation is the factorized distorted wave off-shell impulse approximation, which has been derived and discussed by McCarthy and Weigold.[9] Since it has been discussed in some detail by McCarthy in his talk at this workshop, I will only summarize the more important aspects and include a brief discussion of the case of target state correlations and molecular cross sections.

At fairly high energies the distorted electron waves can be represented to a good approximation by plane waves with wave numbers modified to describe the average real and imaginary potentials felt by the continuum electrons. In this eikonal approximation the (e,2e) cross section on atoms is given by

$$\sigma = \text{const} \frac{k_A k_B}{k_0} |T|^2 R |(2\pi)^{-\frac{3}{2}} \int d^3 r \, \exp(i\underline{q}\cdot\underline{r})(f|g>|^2 \qquad (2)$$

where $|T|^2$ is the half-off-shell Mott scattering cross section, R is the attenuation factor due to absorption of the electron waves, and $|g>$ and $|f>$ are the target and final ionic states respectively. At high enough energies plane waves may be used and R is unity. In the noncoplanar case $|T|^2$ is nearly independent of the angle ϕ and since for large E_0 and small ϵ_i the variation in k_0 is small,

$$\sigma(\phi) \propto |2\pi^{-\frac{3}{2}} \int d^3 r \, \exp(i\underline{q}\cdot\underline{r})(f|g>|^2. \qquad (3)$$

The structure information is all contained in the overlap function $(f|g)$. We can evaluate this by expanding the states $|f)$ and $|g\rangle$ in terms of independent particle configurations $|\alpha\rangle$

$$|g\rangle = \sum_\alpha a_\alpha^{(g)} |\alpha\rangle \qquad (4)$$

The final state is written in terms of a basis consisting of a hole in the orbital ψ_j coupled to target configurations $|\beta\rangle$.

$$|f_r) = \sum_{j\beta} t_{j\beta}^{(f)} [\psi_j^\dagger \times |\beta\rangle]_r$$

$$\equiv n_r^{\frac{1}{2}} \Sigma_{j\beta} t_{j\beta}^{(f)} C_{jr\beta} \psi_j |\beta\rangle \qquad (5)$$

where n_r is the dimension of the representation r of the final state f and the coupling coefficient is a Clebsh-Gordan coefficient. Orthonormality of the independent particle configurations $|\alpha\rangle$ and $|\beta\rangle$ ensures that the overlap function (for a closed shell target) is given by

$$(f_r|g) = n_r^{\frac{1}{2}} \Sigma_{j\alpha} a_\alpha^{(g)} t_{j\alpha}^{(f)} C_{jr\alpha} \psi_j \qquad (6)$$

The factor $n_r^{\frac{1}{2}}$ comes from antisymmetrization.

It is often possible, especially for closed shell targets, to make the independent particle approximation for $|g\rangle$, ($a_0 = 1$, $a_\alpha = 0$ for $\alpha \neq 0$)

$$(f|g) = t_{c0}^{(f)} \psi_c , \qquad (7)$$

where ψ_c is the characteristic single particle orbital and the configuration interaction expansion coefficient of state $|f)$ gives the spectroscopic factor

$$S_c^{(f)} = |t_{c0}^{(f)}|^2 \qquad (8)$$

Since final ion states belonging to the same representation of the point group r can all contain ψ_c, all such states are excited with the relative probability $S_c^{(f)}$. The spectroscopic factors must satisfy the sum rule

$$\Sigma_f S_c^{(f)} = 1 \qquad (9)$$

and the separation energy ϵ_c for the spectroscopic orbital ψ_c is given by

$$\epsilon_c = \sum_f S_c^{(f)} \epsilon_f \qquad (10)$$

That is the single particle eigenvalue is the weighted centroid of the energy eigenvalues of all the states belonging to the representation r containing the characteristic orbital ψ_c.

In this approximation the (e,2e) cross section in the noncoplanar case (where $|T|^2$ is constant) is simply given by

$$\sigma_\phi(q) \propto n_r S_c^{(f)} \sum_{\mu_r} \left| (2\pi)^{-\frac{3}{2}} \int d^3 r \, \exp(i\underline{q}\cdot\underline{r}) \psi_c(\underline{r}) \right|^2$$

$$= n_r S_c^{(f)} \sum_{\mu_r} |\phi_c(q)|^2 \qquad (11)$$

where the incoherent sum is over the projection quantum numbers μ_r of the representation r. The cross section is therefore directly proportional to the square of the momentum space wave function of the characteristic orbital ψ_c weighted by the spectroscopic factor $S_c^{(f)}$. In the coplanar geometry the cross section also depends on $|T|^2$ which depends on the outgoing momenta, increasing rapidly with decreasing angle θ.

When the target state is correlated the sum over j in eq. (6) must in general be carried out, and the cross section will depend a sum of such terms with different characteristic orbitals.

$$\sigma_\phi(q) \propto \left| \sum_{\alpha j} \phi_j(q) a_\alpha^{(g)} t_{j\alpha}^{(f)} C_{j\alpha r} \right|^2 \qquad (12)$$

A particularly simple example of this is He, where the final states are all single particle states and only the target state has correlation effects. Thus the He ground state can be expanded in terms of independent particle configurations (eq. 8) (such as $|g\rangle = a_0 1s^2 + a_1 2p^2 + \ldots$) and we can get excitation of the 2p He$^+$ state as well as the 1s, 2s, etc. states. The ratio of the cross section for exciting the n=2 levels of He will therefore depend on the q distribution of the 2p orbital as well as the 1s orbital of helium. The ratio will therefore depend on q. (I want to remind you that in the THF approximation only $j = j' = c$ and the ratio would be independent of q.)

Let me now turn to molecules where we must generalize the sum μ_r describing the degenerate final states as well as include the sum over initial rotational states. We assume that the molecule is in the ground vibrational state $|0\rangle$. If the Born-Oppenheimer approximation is made we can write the structure wave function as a product of the rotational, vibrational and electronic wave functions. The sum over the initial rotational states is then trivial since the (e,2e) cross section is independent of the total angular momentum I of the molecule [9].

The expression for the cross section can be greatly simplified by applying the closure relation for the final rotational and vibrational states, since these are presently not resolved in the experiments. This gives

$$\sigma \propto |T|^2 \sum_{\mu_r} \langle 0| \int d\Omega (2\pi)^{-3} |\int d^3r \, \exp(i\underline{q}\cdot\underline{r})(F|G\rangle|^2 |0\rangle \qquad (13)$$

where the integral indicated by Dirac brackets is an average over the ground state vibrations and the integral over Ω is the rotational integral. $|G\rangle$ and $|F)$ are the electronic eigenstates of the molecule and ion respectively. Accurate calculations for H_2 [10] show that this average can in that case be replaced by the equilibrium value. I will make this approximation for all molecules.

As before the ground electronic state $|G\rangle$ is expressed as a linear combination of independent-particle basis configurations $|\alpha\rangle$, and the antisymmetrized ion state $|F)$ is written in terms of a basis consisting of a hole in the orbital ψ_j coupled to the target configurations $|\beta\rangle$

$$|F_r) = n_r^{\frac{1}{2}} \sum_{j\beta} t_{j\beta}^{(F)} C_{jr\beta} \psi_j^\dagger |\beta\rangle \qquad (14)$$

The Clebsh-Gordan coefficient of the point group of the molecule ensures that the configurations in the sum (14) all belong to the irreducible representation r. The degeneracy of the representation r is n_r. The subscript r is omitted when this causes no confusion.

For closed shell molecules I will again make the Hartree-Fock approximation for the target which gives

$$(F|G\rangle = n_r^{\frac{1}{2}} \sum_j t_{j0}^{(F)} C_{jr0} \psi_j(\underline{r}\,')|0\rangle = \sum_{j=r} t_{j0}^{(F)} \psi_j(\underline{r}\,') \qquad (15)$$

The Clebsch-Gordon coefficient selects those configurations for which the orbital representation j is the same as the final state representation r. There is a coherent sum over different orbitals that contribute to $|F)$.

The molecular independent particle states $|\alpha\rangle$ which define the basis for the electronic eigenstates $|G\rangle$ and $|F)$ are usually expressed in terms of the LCAO picture. Each molecular orbital $\psi_j(\underline{r}\,',\underline{R}\,')$ in the independent particle determinant $|\alpha\rangle$ is written as a sum of terms centered at each atom centre $\underline{R}_s'$, each term having angular properties given by a spherical harmonic $Y_{\ell m}(\underline{r}\,'-\underline{R}_s')$

$$\psi_j(\underline{r}',R) = \sum_s \psi_s^{(j)}(\underline{r}' - \underline{R}'_s) \tag{16}$$

$$\psi_s^{(j)}(\underline{r}) = \sum_{\ell m} u_\ell^{(s)}(r) Y_{\ell m}(\hat{r}) \tag{17}$$

On making the target Hartree-Fock approximation, the cross section is then given by [9]

$$\sigma \propto S_j^{(F)} n_r |T|^2 \sum_{nm} \int d\underline{q}\, \phi_n^*(\underline{q},R_{nm}) \phi_m(\underline{q},R_{nm}) \exp[i\underline{q}\cdot(\underline{R}_m - \underline{R}_n)], \tag{18}$$

where $S_j^{(F)} = |t_{j0}^{(F)}|^2$ is again the spectroscopic factor for the state to contain the orbital j, n_r the dimension of the representation of the corresponding point group r, R_{nm} the set of equilibrium distances between the nth and mth nucleus, and the momentum space wave function is

$$\phi_n(\underline{q},R) = 2\pi^{-\frac{3}{2}} \int d^3 r \, \exp(i\underline{q}\cdot\underline{r}) \psi_n^j(r,R) \tag{19}$$

where ψ_n^j is the sum of all atomic functions centered on the nucleus m in the LCAO for the characteristic orbital.

Therefore in the noncoplanar case, for which $|T|^2$ is nearly constant, the cross section at different energies should again be simply a function of the recoil momentum q.

I will now discuss how well the experimental results on atoms and molecules are described by this simple picture and how the (e,2e) reaction can be used to yield information on the dynamic structure of atoms and molecules.

EXPERIMENTAL RESULTS

A. ATOMS

Helium offers a test case for the theory since it is the simplest target in the absence of a suitable atomic hydrogen source. In addition, since the final ion eigenstates are single particle which are known exactly, it should be possible to examine details of the helium ground state wave function. Figure 6 shows that the shape of the noncoplanar cross section [9,11] is independent of the energy E over a large range of energies and is very well described by the plane wave theory using a Hartree-Fock wave function. Correlated ground state wave functions have a negligible effect on the shape of the expected momentum distributions.

However, when the overlap between the initial and final state is close to zero, the effects of correlations can become marked. Fig. 7 shows that ratio for excitation of n = 2 states to that of n = 1 depends markedly on q, rising from a value of 0.0074 at small q to the

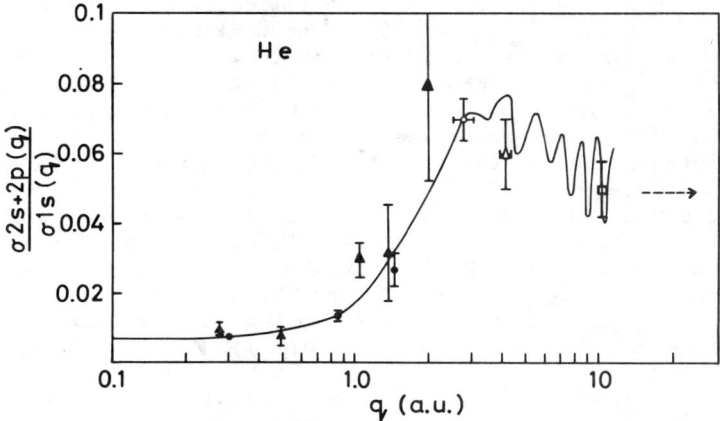

Fig. 7. Ratio of (e,2e) cross sections to He$^+$ n=2 and n=1 states as a function of q. ●, 800 coplanar [12] and ▲ 1200 eV noncoplanar [13] data. O, Δ, ⎯ photoelectron results [14]. The solid curve is the plane wave calculation using the correlated wave function of [15].

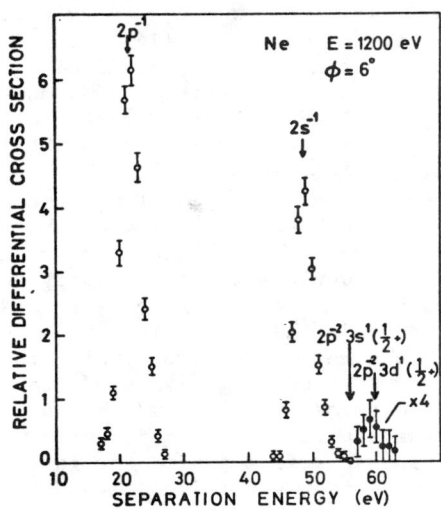

Fig. 8. 1200 eV separation energy spectrum for Ne indicating the main ion states.

order of .08 at a q of 2.0 a.u. The filled circles and triangles are 800 eV coplanar [12] and 1200 eV noncoplanar [13] data respectively. The figure also shows some photoelectron cross sections [14], which at energies above 100 eV are at high q since the photon has very small momentum. One problem with the (γ,e) data is that, due to the difference in separation energy, the cross sections to the different final states are at different q for the same photon energy.

The curve shows the results of the plane wave theory using a high quality correlated wave function [15]. You can see that the agreement is excellent. Hartree-Fock wave functions for He predict excitation of the 2s He^+ state, since the atom and ion states are not completely orthogonal. However in the Hartree-Fock approximation the ratio should be of the order of 0.02 [12] across the entire q region.

It is obvious from the figure that final states with small overlap with the ground state are particularly sensitive to ground state correlations. Further, cross sections for exciting final states in high energy photoelectron spectroscopy (where a plane wave analysis may be justified) are obtained at high q (very close to the nucleus) and these cannot be extrapolated to lower q. I must emphasize that the relative cross section for exciting satellite lines will in general be very q dependent if target state correlations are important.

The other inert gases, Ne, Ar, Kr, and Xe have all been studied at a number of energies in both the coplanar and noncoplanar geometries [6,7,8,9,11,16]. Separation energy spectra [e.g. fig. 8,9, 10] and angular correlations [e.g. figs. 11,12 and 13] show that the ion ground states contain essentially all of the valence np^{-1} strength but that the ns^{-1} strength is split among a number of ion eigenstates, especially for Ar, Kr, and Xe. In fact for the latter two elements the ion state whose dominant configuration is given as ns^{-1} contains only one third of the ns hole strength.

The experiments do not at present measure absolute spectroscopic factors, but they do measure the ratios of cross sections to different final states. Since the np hole is not significantly split we can give it a spectroscopic strength of unity and normalize our ns spectroscopic factors relative to this. If this is done the sum of the ns^{-1} strengths is in the range of 0.7±0.1 for argon, krypton and xenon, for energies below 1000 volts. This shows that absorbtion is more serious for the more tightly bound states than for the outer p states. If distortions are included in the calculation the spectroscopic strengths are found to sum to unity within experimental error ($\leq$ 0.1). We have also measured the ratio of exciting the $5p_{3/2}$ to $5p_{1/2}$ Xe ion states and found it to be 2.0±011, in agreement with the expected value of 2. Hood has obtained a similar result at the University of British Columbia.

In table 1 I have compared the relative ns^{-1} strengths (normalized to unity) observed in (e,2e) and in (γ,e) at a number of energies. The (e,2e) data (at q $\approx$ 0) are in agreement with each other over the entire range of energies 200 eV to 2500 eV. However, except for Ne, they are very different from the high q XPS results.

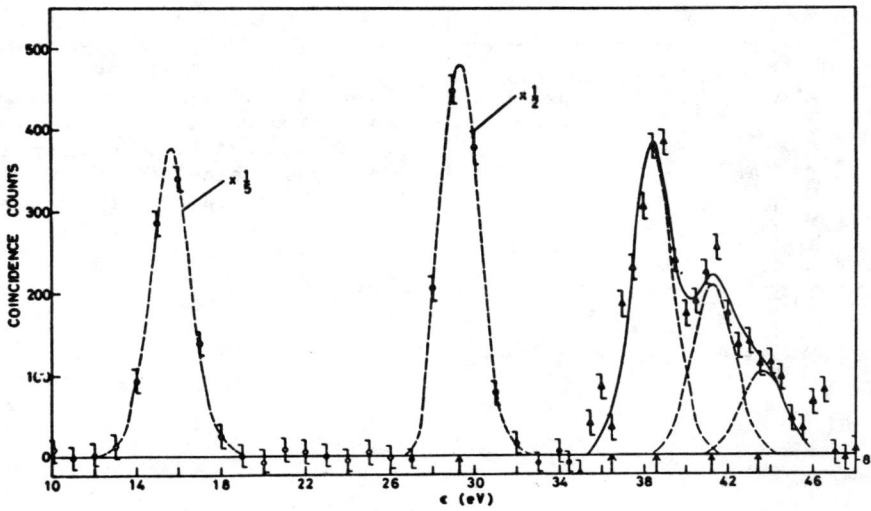

Fig. 9. 400 eV separation energy spectrum for argon.

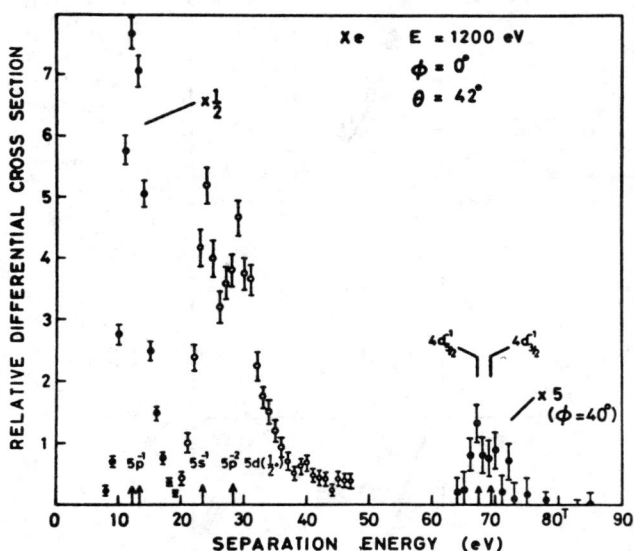

Fig. 10. 1200 eV separation energy spectrum for Xe with the main ion state configuration as indicated.

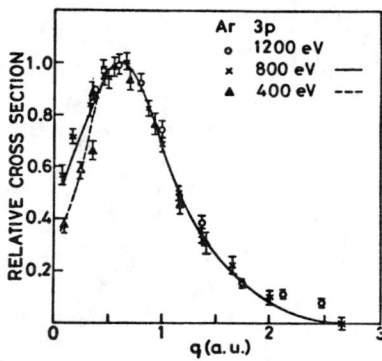

Fig. 11. Noncoplanar symmetric (e,2e) cross section for the argon ground state transition plotted as a function of q. The finite angular resolution has been folded into the plane wave calculations using Hartree-Fock 3p wave functions.

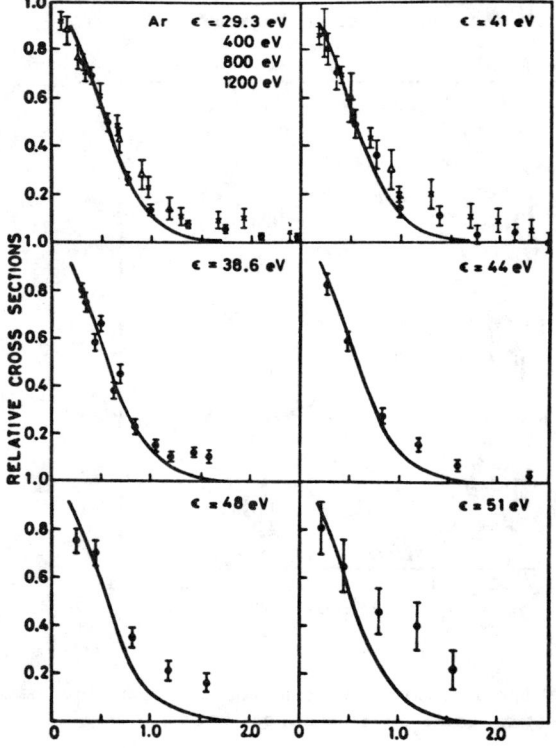

Fig. 12. Noncoplanar symmetric (e,2e) angular correlations for (e,2e) on argon leading to final states with the indicated separation energies. The curves are the Hartree-Fock 3s momentum distributions.

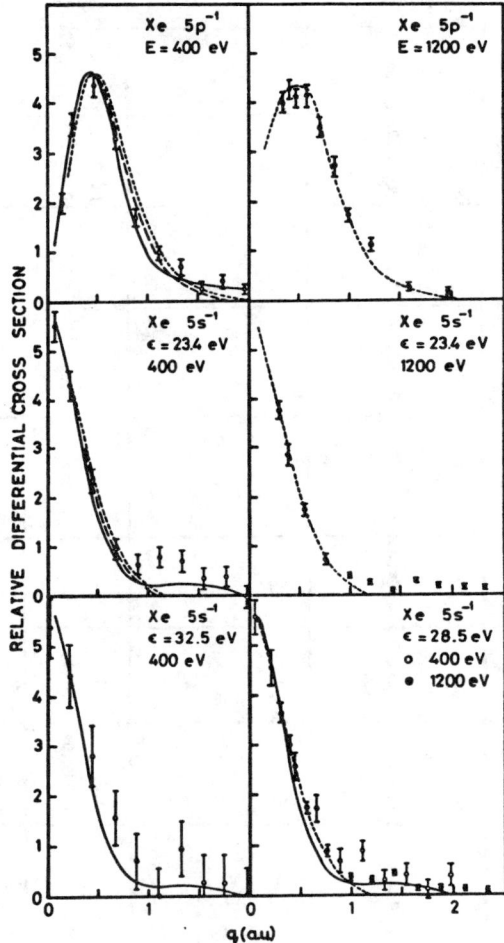

Fig. 13. Noncoplanar symmetric Xe (e,2e) angular correlations for different separation energies ϵ. The solid and dotted curves are the phase distorted [9] and plane wave calculations respectively, the dashed curve being the eikonal approximation.

Table 1. Comparison of (e,2e) and XPS spectroscopic strengths normalized to unity for the valence s hole states of inert gases.

Hole State	Separation energy (ε) in eV			Dominant Ion state configuration	Relative Spectroscopic Strengths							
					(e,2e)					XPS		
	Optical	(e,2e)	XPS		200eV[a]	400eV[a]	800eV[a]	1200eV[a]	2500eV[b]	151eV	1254eV	1487eV
Ar3s⁻¹	29.25	29.3±1	29.3±1	$3s^{-1}$	0.57	0.58	0.59	0.52	0.66	0.81	0.82	0.80
	38.58	38.6±1	38.6±1	$3p^{-2}3d$		0.23±.02				0.12±.02	0.12±.02	0.15±.02
	41.20	41.2±2	41.2±1	$3p^{-2}4d$	0.43±.04	0.13±.02	0.41±.03	0.40±.02	0.34±.07	0.07±.03	0.06±.03	0.05±.03
		43.4±1	43.4±1			0.06±.02		0.08±.01				
		>44										
Ne2s⁻¹	48.5	48.5	48.5	$2s^{-1}$		0.98		0.96		0.95±.02		
	55.9	60	59.5	$2p^{-2}3s^1$		-		<0.01		<0.04		
	59.88	>60		$2p^{-2}3d^1$		0.02±.02		0.04±.01		<0.02		
			88	$2s^{-1}2p^{-1}3p^1$		<0.05				0.04±.02		
Kr4s⁻¹	27.51	27.5±1	27.5±1	$4s^{-1}$	0.29	0.31			0.37		0.80	0.76
	33.94	Peak at 34	34±1	$4p^{-2}4d^1$	0.71±.04	0.69±.04			0.63±.15		0.20±.03	0.24±.02
Xe5s⁻¹	23.40	23.3±1	23.4±1	$5s^{-1}$	0.35	0.34		0.35			0.65-.92	0.58-.90
	27.51	Peak at ≈28.1	27.6±.2	$5p^{-2}6s^1$	0.65±.03	0.66±.03		0.65±.04			0.05-.10	0.04-.10
	28.15		29.0	$5p^{-2}5d^1$							0 - 0.3	0 - 0.4

(a) (e,2e) data from Flinders; (b) Frascati data.

It is interesting to note that no excitation is observed of the $np^{-1}(n+1)s(\frac{1}{2}+)$ states, whereas the $np^{-2}(n')d(\frac{1}{2}+)$ ion eigenstates are strongly excited, suggesting that quadrupole core couplings play an important role in the correlations.

We find that the shape of the noncoplanar symmetric 400 eV cross sections for the large Z atom Xe (fig. 13) are poorly described by plane wave theory but are quite well described by a phase distortion model [9]. At 1200 eV the plane wave theory fits the shapes quite well. However the $4d^{-1}$ angular correlation is very poorly described by the plane wave theory [fig. 14]. Clearly plane waves are inadequate for the more tightly bound inner state even at 1200 eV.

Cliff Noble at Flinders University is at present carrying out calculations on these cross sections using fully distorted electron waves.

B. MOLECULES

Molecular hydrogen provides us with a test molecule in the same way that helium provides a test case for atoms. We have measured the momentum distribution for the ground state transition for H_2 and D_2 very accurately at 300, 400, 600, 800 and 1200 eV [10]. The data are independent of energy and are in agreement with the plane wave calculations if a correlated H_2 wave function is used [fig. 15]. Further the results for H_2 and D_2 are indistinguishable, as expected from the Born-Oppenheimer approximation.

However, as well as the ground state $1s\sigma_g$ transition a small amount of excitation is observed to the $2p\sigma_u$, $2p\pi_u$, and $2s\sigma_g$ states. The ratio of cross sections with separation energies in the range 30 - 45 eV to that of the $1s\sigma_g$ transition depends strongly on q and also on ε. In order to analyze this data, which we are presently doing, it is necessary to use a detailed correlated H_2 ground state wave function.

The first polyatomic molecule we studied was methane. Fig. 17 shows the separation energy spectra for the isoelectronic molecules CH_4 [18], NH_3 [19], and H_2O [20], all showing considerable amount of splitting in the deepest valence orbitals. Fig. 18 shows the observed angular correlations. In the case of CH_4 and H_2O the plane wave cross sections have been normalized at only one point, the relative cross section ratios being measured experimentally. For NH_3, the calculated cross sections have been adjusted to give the best fit. Except for NH_3, the orbital wave functions of Snyder and Basch were used. The dashed curve for NH_3 uses an STO wave function, the solid curve is obtained if a much better wave function is used [19]. It is interesting to note that the ground state transition for all of these molecules has too many low q components compared to the calculated cross sections. This is particularly severe for water, and it may be necessary to do a full overlap calculation with vibrational averaging in this case. The 400 and 1200 eV H_2O data are in excellent agreement with each other, so the momentum distribution is independent of energy, as it has to be if the plane wave model is correct.

102

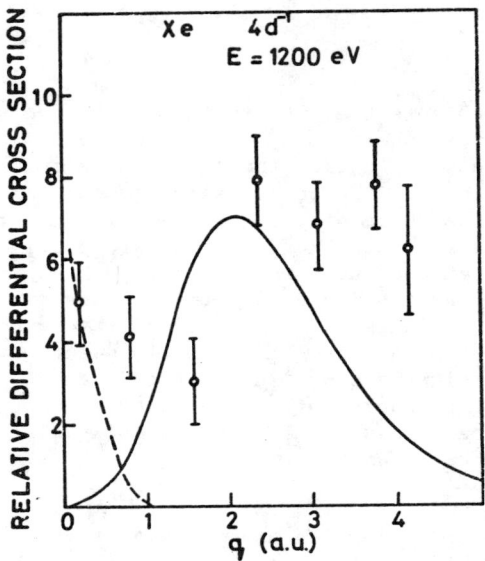

Fig. 14. The Xe $4d^{-1}$ momentum distribution. The dashed and solid curves are the $5s^{-1}$ and $4d^{-1}$ plane wave calculations respectively.

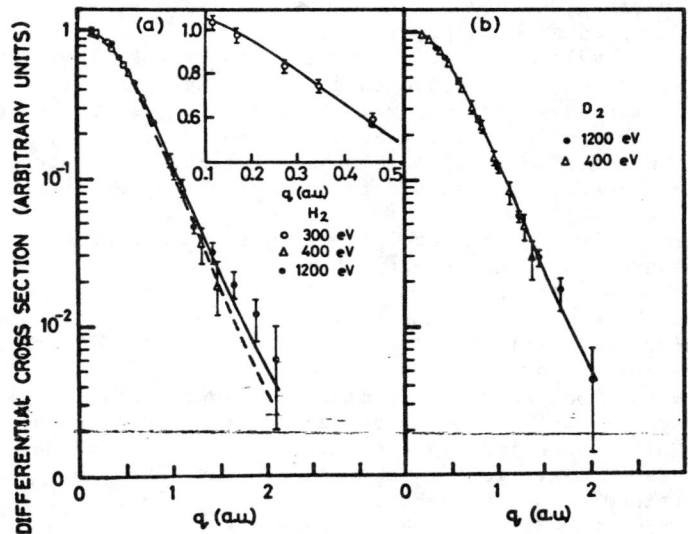

Fig. 15. The noncoplanar (e,2e) ground state H_2 and D_2 cross sections plotted as a function of q. The solid curve is a plane wave calculation using a CI wave function whereas the dashed curve uses the orbital wave function of Snyder and Basch [17].

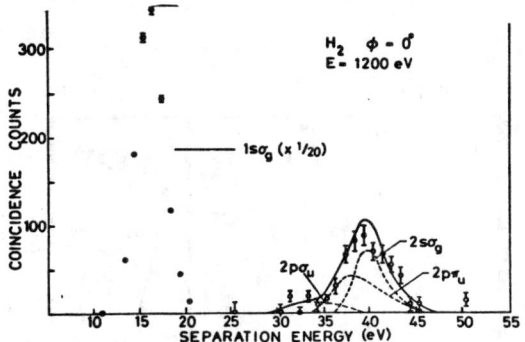

Fig. 16. The differential cross section for the 1200 eV (e,2e) reaction on H_2 at $\phi = 0°$ plotted as a function of ϵ. The final ion states are as indicated.

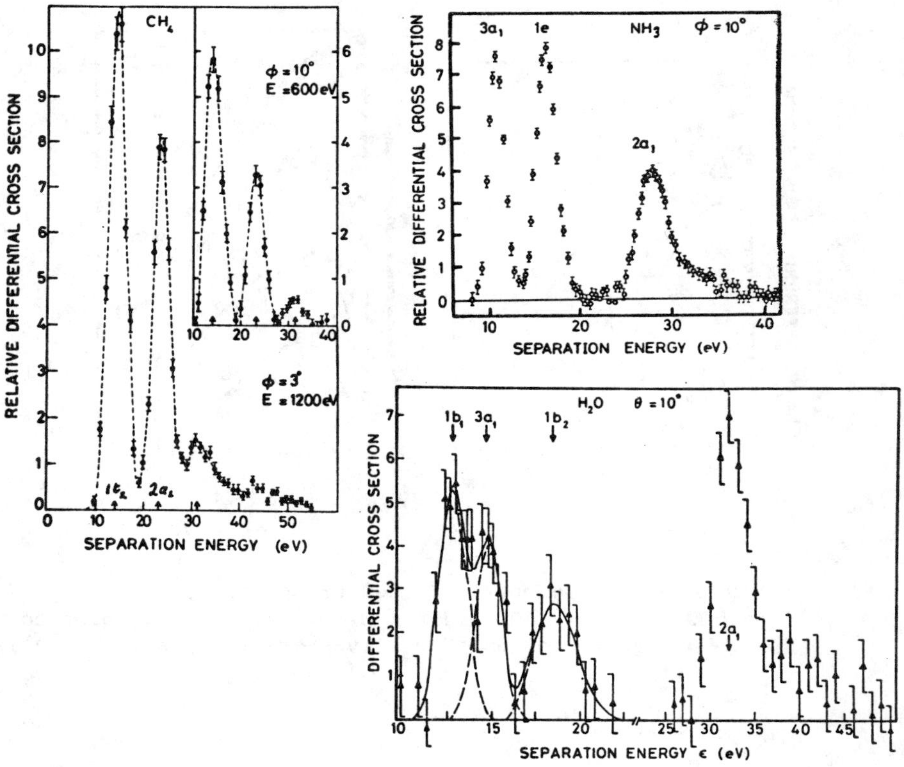

Fig. 17. Separation energy spectra for CH_4, H_2O, and NH_3.

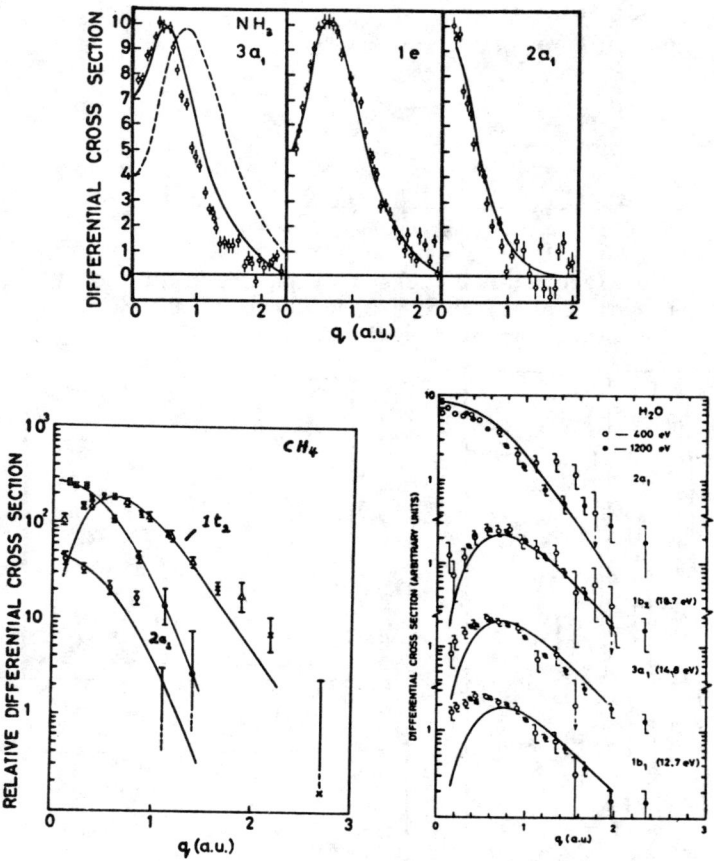

Fig. 18. Noncoplanar symmetric (e,2e) cross sections for CH_4 (600 eV), NH_3 (400 eV) and H_2O (400, 1000 eV) plotted as a function of q. The curves are the plane wave calculations.

Fig. 19 shows separation energy spectra in the valence region for the isoelectronic diatomic molecules CO and N_2 [21,22]. Considerable configuration interaction effects are present, not all due to the deepest valence orbital. Fig. 20 shows the observed angular correlations compared with plane wave calculations. The agreement is excellent both in magnitude and shape. The figure also shows the angular correlations for various smaller contributions (satellite) in the separation energy spectra, together with the shapes expected for different orbitals.

The separation energy spectra and momentum distributions obtained for ethane are shown in figures 21 and 22.[23] The plane wave calculations are in good agreement with the data. There has been some controversy about the symmetry of the ground state of the C_2H_6 ion, since the $3a_{1g}$ and $1e_g$ states overlap. Our (e,2e) data show quite clearly the ground state of $C_2H_6^+$ belongs to the $3a_{1g}$ orbital. Similarly Hood, Hamnett and Brion recently used the (e,2e) reaction for orbital assignment in formaldehyde.[24]

CONCLUDING REMARKS

I have tried to give an outline of what has been achieved to date in determining momentum and structure information on atoms and molecules using the symmetric (e,2e) reaction. Although I have largely omitted the considerable body of coplanar symmetric work, the results presented show that the technique is unique in investigating atomic and molecular structure. The noncoplanar arrangement is particularly suitable for obtaining structure information since it is insensitive to distortion. On the other hand, the coplanar arrangement is particularly sensitive to distortions, especially in the forward directions ($\theta < 45°$).

Of course we have only just begun. For instance there is a great need to measure absolute cross sections. Such an experimental program has begun at FOM in Amsterdam, a collaboration between F. deHeer, B. Van Wingerden and myself. This experiment was also designed to make accurate energy sharing measurements. The suggestion of P. Roos earlier in his (p,2p) talk to track the q=0 point through its kinematical region applies also to (e,2e). This should give valuable information on the factorization approximation.

Thus far we have also only looked at closed shell targets in both atoms and molecules. Plans are afoot both at Frascati and Flinders to study alkalis and other open shell targets (e.g. NO). Last, but not least, we would obviously like to apply the (e,2e) technique to study surfaces. Again several laboratories are pursuing this, and we will hear from N. Avery on his work in this area.

New equipment incorporating position sensitive electron detectors is being constructed at Flinders. This should significantly improve the rate of taking data at high energy resolution, making possible the study of much more complicated and chemically interesting molecules.

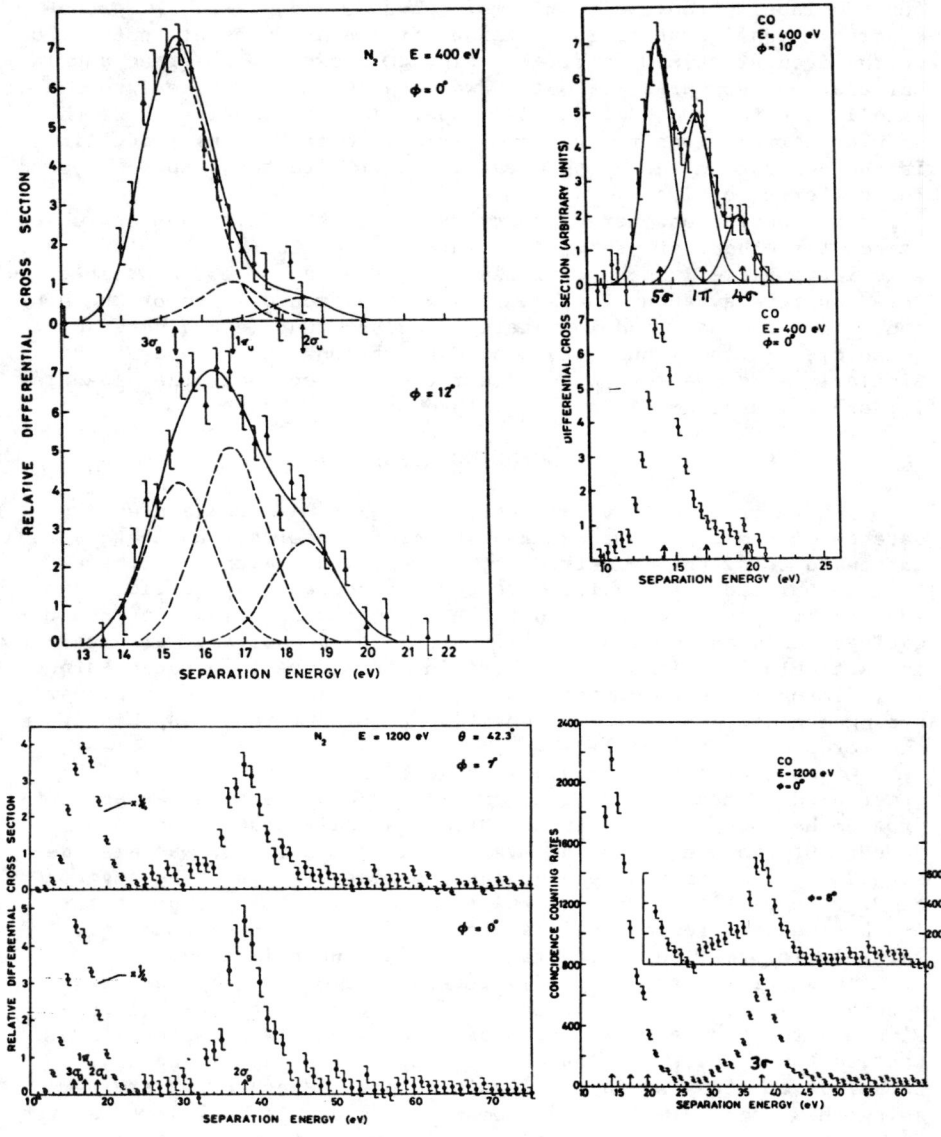

Fig. 19. Separation energy spectra for CO and N_2 in the valence region with the vertical ionization energies of the dominant peaks of various orbitals indicated [21,22].

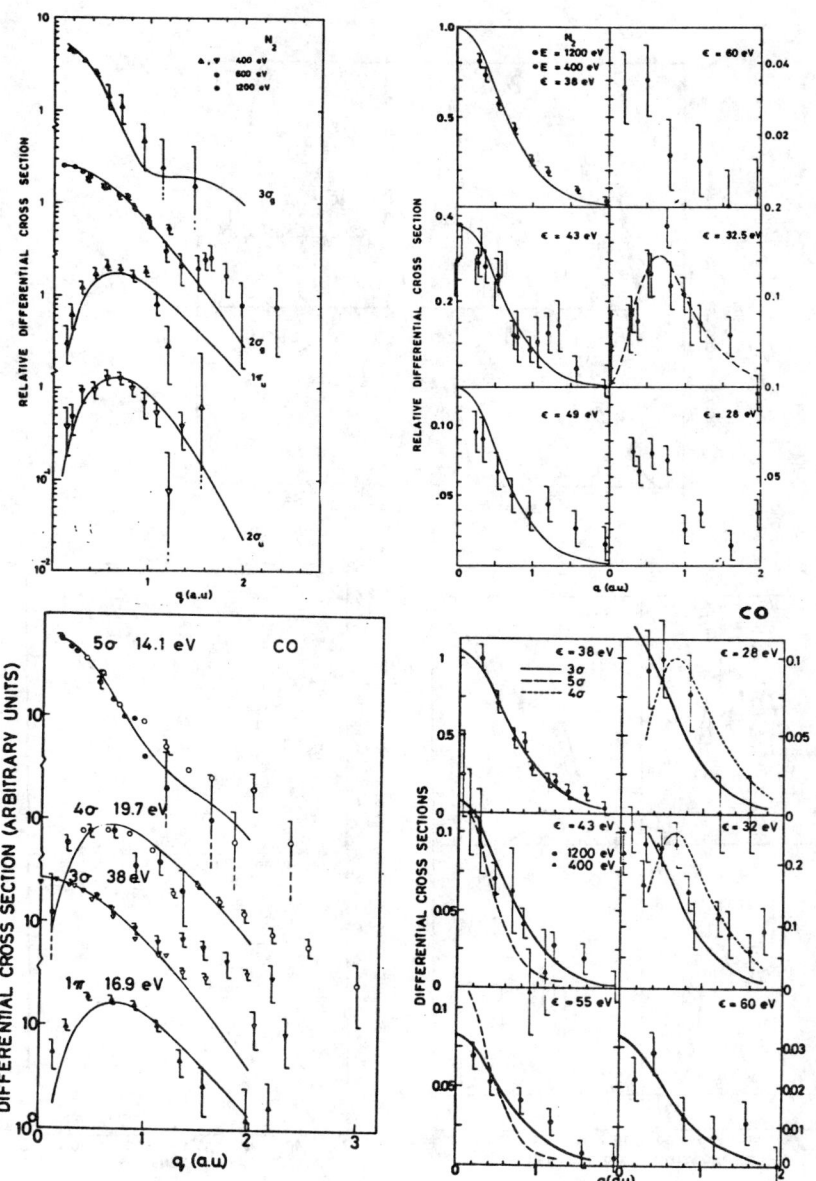

Fig. 20. Noncoplanar symmetric (e,2e) cross sections for CO and N_2 at 400 eV and 1200 eV compared in magnitude and shape with the plane wave calculations using the orbital wave functions of Snyder and Basch [17]. Also shown on the linear plots are the angular correlations observed for ?I peaks in the separation energy spectra [21,22].

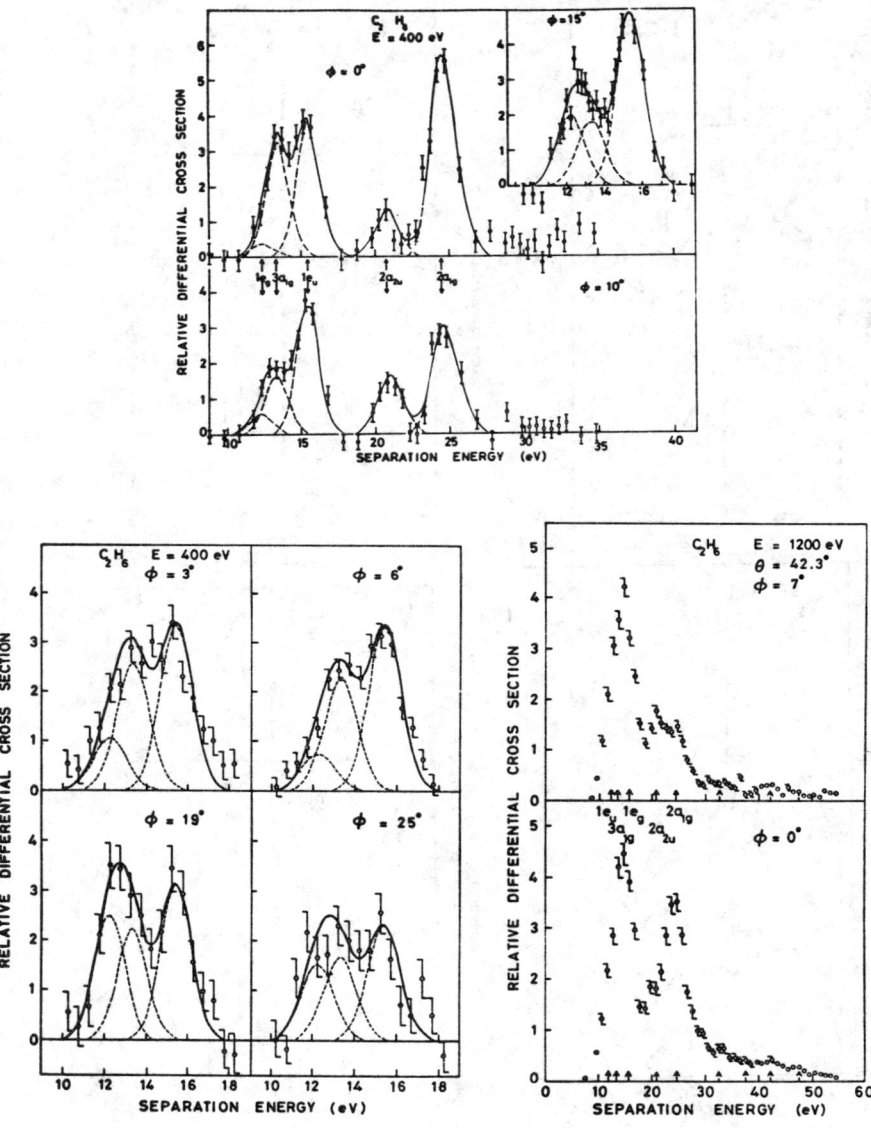

Fig. 21. Separation energy spectra for C_2H_6 at a number of angles [23].

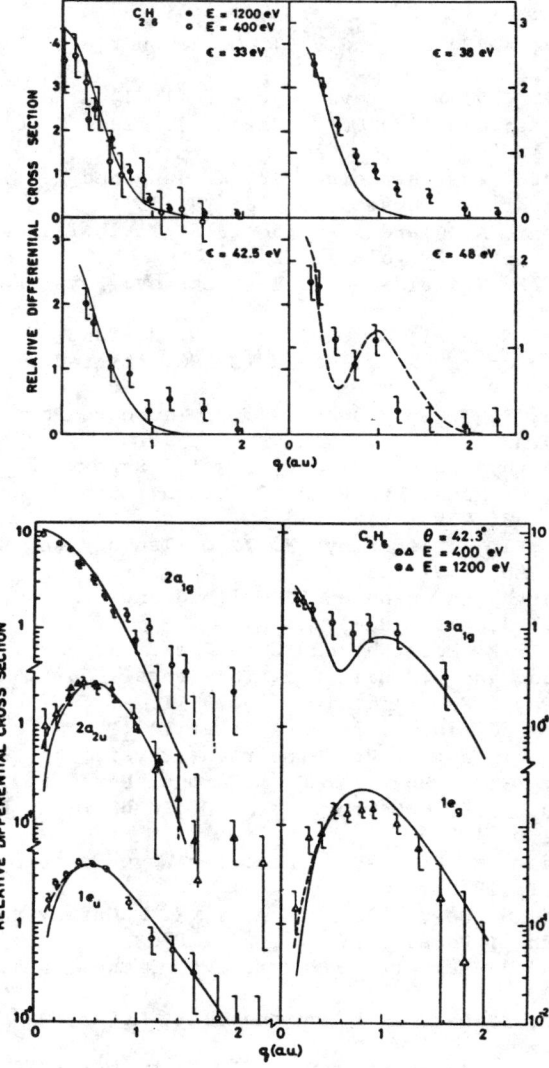

Fig. 22. Noncoplanar symmetric (e,2e) cross section for C_2H_6 at 400 eV and 1200 eV compared in shape and magnitude with pland wave calculations using the orbital wave functions of Snyder and Basch [17].

REFERENCES

1. E. Weigold, A. Ugbabe and P. J. O. Teubner, Phys. Rev. Lett. $\underline{35}$, 209 (1975).
2. M. Inokuti, Rev. Mod. Phys. 43, 297 (1971).
3. M. J. van der Wiel and C. E. Brion, J. Elec. Spectrosc. I 309, (1973); ibid 443.
4. H. Ehrhardt, J. H. Herselbacher, K. Jung and K. Willmann, Case Studies in Atomic Physics $\underline{2}$, 618 (1971).
5. R. Camilloni, A. Giardini-Guidoni, R. Tiribelli and G. Stefani, Phys. Rev. Lett. $\underline{29}$, 618 (1972).
6. A. Ugbabe, E. Weigold and I. E. McCarthy, Phys. Rev. A$\underline{11}$, 566 (1975).
7. A. Giardini-Guidoni, R. Camilloni, G. Stefani, R. Tiribelli, D. Vinciguerra and E. Weigold, IX ICPEAC, Abstracts of Papers, 490 (1975).
8. E. Weigold, S. T. Hood and P. J. O. Teubner, Phys. Rev. Lett. $\underline{30}$, 475 (1973).
9. I. E. McCarthy and E. Weigold, Physics Reports $\underline{27C}$, 275 (1976).
10. S. Dey, I. E. McCarthy, P. J. O. Teubner and E. Weigold, Phys. Rev. Lett. $\underline{33}$, 459 (1974).
11. S. T. Hood, I. E. McCarthy, P. J. O. Teubner and E. Weigold, Phys. Rev. A$\underline{8}$, 2494 (1973).
12. I. E. McCarthy, A. Ugbabe, E. Weigold and P. J. O. Teubner, Phys. Rev. Lett. $\underline{33}$, 459 (1974).
13. A. Dixon, I. E. McCarthy and E. Weigold, J. Phys. B$\underline{9}$, L195 (1976).
14. M. O. Krause and F. Wuilleumier, J. Phys. B$\underline{5}$, L143 (1972); T. A. Carlson, Phys. Rev. $\underline{156}$, 142 (1967); T. A. Carlson, M. O. Krause and W. E. Moddeman, J. de Phys. (Paris) $\underline{10}$, C4-76 (1971).
15. C. J. Joachain and R. Vanderpoorten, Physica $\underline{46}$, 333 (1970).
16. E. Weigold, S. T. Hood, I. E. McCarthy, Phys. Rev. A$\underline{11}$, 566 (1975); S. T. Hood, I. E. McCarthy, P. J. O. Teubner and E. Weigold, Phys. Rev. A$\underline{9}$, 260 (1974).
17. L. C. Snyder and H. Basch, Molecular Wave Functions and Properties (Wiley, N.Y. 1972).
18. E. Weigold, S. Dey, A. J. Dixon, I. E. McCarthy and P. J. O. Teubner, Chem. Phys. Lett. $\underline{41}$, 21 (1976).
19. S. T. Hood, A. Hamnett and C. E. Brion, Chem. Phys. Lett. $\underline{39}$, 252 (1976).
20. E. Weigold, S. Dey, A. Dixon and I. E. McCarthy (Chem. Physics, in press).
21. S. Dey, A. J. Dixon, K. R. Lassey, I. E. McCarthy, P. J. O. Teubner, E. Weigold, P. Bagus and E. K. Viinikka (Phys. Rev. A, in press).
22. E. Weigold, S. Dey, A. J. Dixon, I. E. McCarthy, K. R. Lassey and P. J. O. Teubner, J. Elec. Spectrosc. $\underline{10}$, 177 (1977).
23. S. Dey, A. J. Dixon, I. E. McCarthy and E. Weigold, J. Elec. Spectrosc. $\underline{9}$, 397 (1976).
24. S. T. Hood, A. Hamnett and C. E. Brion, Chem. Phys. Lett. $\underline{41}$, 428 (1976).

NUCLEON KNOCKOUT: REACTION MECHANISMS

Edward F. Redish*
Department of Physics and Astronomy
University of Maryland, College Park, Maryland 20742 USA

INTRODUCTION

In the talks of Professors Walker and Roos earlier in this conference, we learned that the (p,2p) reaction at reasonably high energies can be conceived of as a probe of the shell-model wave functions of the struck nucleon. This expectation arises directly from the "pole-dominated" or one-step plane wave picture of the knockout. Unfortunately, nature is not always so agreeable as to have things happen in exactly the way we would like them to happen, especially when a many-body problem is involved. I will consider two questions. First: what actually happens in nucleon knockout? Do the deviations from the simplest plane wave picture interfere with an extraction of the wave function? Second: I will ask whether or not our desire to measure the bound state wave functions blinds us to the possibility of obtaining other interesting information from this reaction, or from finding exciting new processes taking place.

THE PLANE WAVE IMPULSE APPROXIMATION

Let's begin by looking carefully at the amplitude for the plane wave impulse approximation. Qualitatively, the idea is that the incoming nucleon interacts perturbatively with one of the nucleons in the nucleus and knocks it out. This is shown schematically in Fig. 1. If we label the struck nucleon by 1 and the projectile by 0, then the Hamiltonians in the initial and final states are

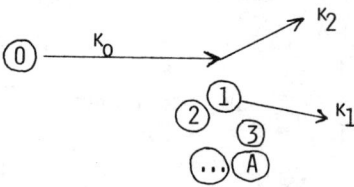

Fig. 1. The PWIA

$$H_I = T_0 + H_A = T_0 + \sum_{i=1}^{A} T_i + \sum_{i<j=1}^{A} V_{ij} \qquad (1)$$

$$H_F = T_0 + T_1 + H_{A-1} = T_0 + T_1 + \sum_{i=2}^{A} T_i + \sum_{i<j=2}^{A} V_{ij} \qquad (2)$$

The operator T_i is the kinetic energy of particle i and V_{ij} is the interaction potential of the pair ij. The initial state consists

*Supported by U. S. Energy Research and Development Administration.

of an incoming free particle on a bound nucleus, and satisfies

$$H_I |k_0 \Phi_A\rangle = (\frac{\hbar^2 k_0^2}{2M} + E_A) |k_0 \Phi_A\rangle \ . \tag{3}$$

The final state has two outgoing free particles on a bound (perhaps highly excited) residual nucleus and satisfies

$$H_F |k_1 k_2 \Phi_{A-1}\rangle = (\frac{\hbar^2 k_1^2}{2M} + \frac{\hbar^2 k_2^2}{2M} + E_{A-1}) |k_1 k_2 \Phi_{A-1}\rangle \ . \tag{4}$$

Assuming some effective transition operator, V_{eff}, we get the matrix element

$$M_{FI}^{(PW)} = \langle k_1 k_2 \Phi_{A-1} | V_{eff} | k_0 \Phi_A \rangle \tag{5}$$

The first uncertain question is what one should choose as the effective interaction V_{eff}. In nuclear physics, the answer to this is not as simple as in molecular or atomic physics where the electron-electron interaction is well known. The nucleon-nucleon force is very strong, but has a strong short range repulsion almost precisely balanced by a longer range attraction. This almost precise cancellation results in weakly bound systems extending over regions large compared to the range of the force. This effect is shown in Figure 2 (from ref. 1) which compares the two-nucleon bound state wave functions with the two-hydrogen atom bound state in natural units. For heavier nuclei the effect is not so dramatic but still exists. In the interiors of heavy nuclei, densities

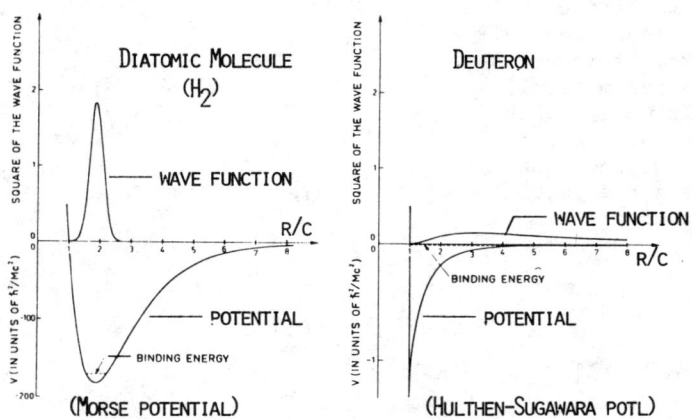

Fig. 2. Comparison of atomic and nuclear bound states.

saturate at about 0.17 nucleons/fm^3. This corresponds to an average separation distance of $(1/0.17)^{1/3} \sim 1.8$ fm. For light nuclei and in the nuclear surface (where many reactions take place), the spacing is even larger than this. If the radius of the nucleon may be taken as the core radius (about 1/2 fm), the nucleus looks like a <u>dilute</u> system to a nuclear probe. We may, therefore, consider a perturbation-like expansion where the probe interacts with one nucleon at a time.

We're not allowed, however, to treat the force perturbatively. Although the average force is weak, locally it's very strong. The relative two particle wave function is distorted very severely at short distances. Therefore, we include the effect of the force between a given interacting pair to all orders. We do this by means of the series

$$V_{eff} = V_{01} + V_{01} G_A V_{01} + V_{01} G_A V_{01} G_A V_{01} + \cdots \qquad (6)$$

where

$$G_A = \frac{1}{E^+ - T_0 - T_1 - H_{A-1}} \,. \qquad (7)$$

In between the interactions the rest of the nucleons are allowed to go on their merry way. The series may be summed by the integral equation

$$V_{eff} = t_{01} \qquad (8)$$

$$t_{01} = V_{01} + V_{01} G_A t_{01} \,. \qquad (9)$$

We note that this operator is not exactly the transition operator for nucleon-nucleon scattering. That operator is given by

$$\hat{t}_{01} = V_{01} + V_{01} \hat{G}_{01} \hat{t}_{01} \qquad (10)$$

with

$$\hat{G}_{01} = \frac{1}{E^+ - T_{01}^{rel}} \,. \qquad (11)$$

The operator T_{01}^{rel} is the kinetic energy of the relative motion of the pair 01. The Green functions in Eqs. (9) and (10) are different, the one in two-nucleon scattering propagating only the relative motion of the pair. By decomposing the total kinetic energy of the pair into relative and CM by

$$T_0 + T_1 = T_{01}^{rel} + T_{01}^{cm} , \qquad (12)$$

the operator we have may be related to the two-nucleon transition operator at a shifted energy:

$$t_{01}(E) = \hat{t}_{01}(E - T_{01}^{cm} - H_{A-1}) . \qquad (13)$$

Since the potential V_{01} involve only the relative coordinate of the pair of particles and not their CM coordinate, this T matrix is diagonal in the CM momentum of the pair, i.e., the interaction conserves momentum. As we know the initial momentum of one of the nucleons and the final momenta of both, the fourth momentum may be deduced. The initial and final relative momenta of the pair may then be calculated. These are plotted in Figure 3 for the particular case of 180 MeV incident protons with the final protons observed in coplanar symmetric geometry[2]. We observe that the initial and final momenta appearing in the scattering amplitude never agree in magnitude for any angle, the difference being most severe at the forward angles. The scattering is, therefore, necessarily an off-energy-shell amplitude which cannot be taken from two-body experiments. (The final relative momentum is labelled p_{on} since the energy shift in the argument of the T matrix shifts the energy appearing in the Green function to equal the final relative kinetic energy of the pair.)

This causes little difficulty in (e,2e) experiments as the force is known exactly, but this is not the case in nuclear physics. The degree of uncertainty introduced by the appearance of an off-shell amplitude is not as great as one might first expect if one doesn't do wild things. The reason for this will be discussed by Dr. Stephenson in his talk. One extremely interesting point is that considerable effort has been made in nuclear physics to obtain information on the nucleon-nucleon force. In general, the full T matrix as a function of all three of its variables is needed to determine the potential exactly. We only know it for on shell

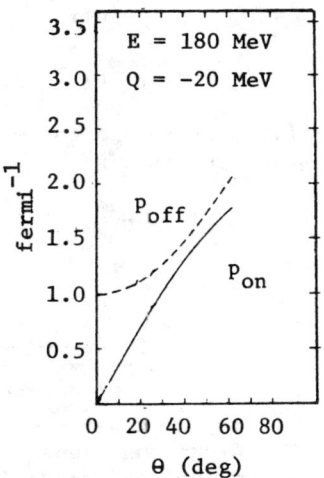

Fig. 3. On and off shell momenta for coplanar symmetric geometry.

values (the heavy line in Fig. 4) and at the bound state pole if the wave function is known. This pole in energy is the plane $E = E_B$ shown in Fig. 4. If constraints could be put on the force from knockout reactions, they would be at least as valuable as the spectroscopic information usually sought.

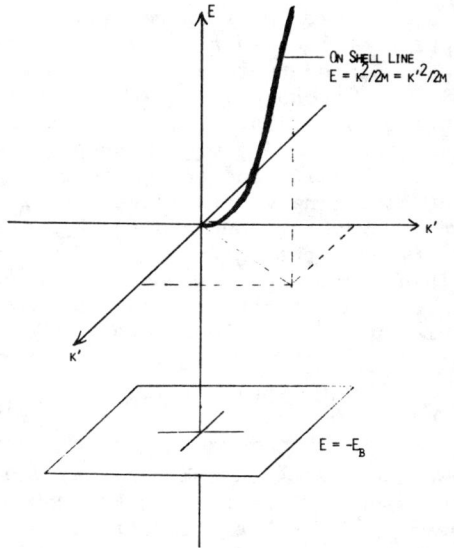

Fig. 4. Variables of the fully-off-shell T matrix.

THE DISTORTED WAVE IMPULSE APPROXIMATION

In nuclear physics, the interaction is quite strong and, although the nucleus is reasonably dilute, it is not in general transparent. Around 100 MeV, for example, the proton-proton total nuclear cross section is about 25 mb. This corresponds to a disk with a 1 fm radius. The mean free path of a proton through nuclear matter should then be something like $L = 1/\rho\sigma$ (the standard formula from classical statistical mechanics). This yields a mean free path on the order of 2 fm. Since even light nuclei have radii of 2 fm or more, we must conclude that our simple plane wave arguments should not hold except at the very edge of the nucleus. In the interior, much scattering will occur and we do well to notice that the mean free path is comparable both to the range of the strong part of the force and to the nuclear spacing. This tells us that things may be uncomfortably more complicated than we had hoped and that it is possible that even our two-body effective interaction may have to be constructed as a many-body operator.

What the nuclear physicist does to get around these difficulties is to say that lots of different things can happen, but when we look at a particular reaction (such as a knockout scattering to a well defined region of the three-body phase space), that it is unlikely to happen and can be treated perturbatively. The presence of the other reactions are included by absorbing (attenuating) the incident beam as it passes through the nucleus. Another way of saying this is that we try to take into account the most important processes as carefully as we can. Since every reaction leads to an absorption of the elastic beam and, therefore, to elastic scattering by shadow scattering, we consider elastic scattering to be the most important channel. We must treat elastic scattering of the projectile carefully. This suggest the following view of the reaction: the incoming proton propagates towards the nucleus, enters the nuclear field, and begins to refract and attenuate. Then the strong scattering occurs which knocks out one of the bound nucleons. Both then propagate outward through the nucleus, refracting and attenuating as they go.

This picture corresponds to the well known distorted wave approximation already discussed at this meeting by many speakers. The relevant matrix element is

$$M_{fi}^{(DW)} = \langle k_1^{(-)} k_2^{(-)} \Phi_{A-1} | t | k_0^{(+)} \Phi_A \rangle \quad (14)$$

We can now again ask the question: what effective interaction should we use to produce the knockout? The simplest thing to take is the same effective interaction we used before. Now things become a bit more complicated. Our effective interaction has operators in its energy variable to remove the CM energy of the interacting pair and the energy binding the struck particle to the core. Opening up Eq. (14) by introducing the momentum representation yields the gory mess displayed below:

$$M_{fi}^{(DW)} = \int dq_0 dq_1 dq_2 dq_3 \chi_{k_1}^{(-)}(q_1)^* \chi_{k_2}^{(-)}(q_2)^*$$

$$\times \langle \frac{q_1-q_2}{2} | t(W) | \frac{q_0-q_3}{2} \rangle \delta(q_1+q_2-q_0-q_3) \chi^{(+)}(q_0) F(q_3) \quad (15)$$

where

$$W = E - (q_0+q_3)^2/4m - Q_{A,A-1} \quad (16)$$

$$F(q) = \langle q \, \Phi_{A-1} | \Phi_A \rangle . \quad (17)$$

We observe two features that make this integral difficult to calculate. First, the angular integrals are hard to do because of the presence of three continuum particles. Making partial wave expansions for each of these scattering wave functions leads to on the order of L^3 integrals, where L is the maximum angular momentum required. Second, the effective interaction is non-local and energy dependent. The whole procedure would be simplified considerably if the T matrix could be extracted from the integral and replaced by some "effective" T matrix. This would reduce Eq. (15) to the simpler form

$$M_{fi}^{(fac)} = t_{eff} \int dr \, \chi_{k_1}^{(-)}(r)^* \chi_{k_2}^{(-)}(r)^* \chi_{k_0}^{(+)}(r) F(r). \tag{18}$$

Roos discussed such factorizations a bit, and Stephenson and Koshel will have more to say about them tomorrow. I will, therefore, only note that there is nothing in the structure of the matrix element which would tend to replace the fully-off-shell matrix element in the DWIA integral by an effective T matrix which is on-shell. About the closest one might hope to get is the half-shell T matrix obtained in the PWIA[3]. One obtains this if one assumes that the spatial fourier components of the elastic waves have a narrow band width about the plane wave. This would be the case, for example, if the straight line paths and fairly weak absorption sufficed.

As we've seen from Roos's talk, the DWIA works quite well, even in factorized form, at energies above 150 MeV. The calculations seem to get a bit worse as we descend in energy. One should remember here that one is necessarily dealing with three-body final states. Our experience with three-body problems warns us that a good fit to the shape of an experimental distribution does not guarantee that the reaction mechanism is being correctly described. Shapes in both the QFS and FSI region can be accurately predicted by an approximation whose normalization is off by a factor of 10 or more from the correct value.

SOME FAILURES

I would now like to mention two results which are difficult to explain within the context of the distorted wave reaction mechanism. The first involves the excitation of forbidden states, the second the energy dependence of the spectroscopic factors discussed by Roos.

The first experiment is $^{12}C(p,2p)^{11}B$ at 50 MeV. The coplanar symmetric angular distributions and a typical summed energy spectrum[4] are shown in Figure 5. We observe that the ground state is strongly excited, and the first four excited states are also seen with comparable strength. The creates a serious problem of interpretation. Fig. 6 shows the ordering of the levels of the nuclear shell model. Carbon fills the lowest s and p shells. We would therefore expect to be able to knock out a $p_{3/2}$ proton. (The $s_{1/2}$ proton is too deeply bound to be knocked out at these energies and angles.) This accounts for the $3/2^-$ ground state, but where did the other states come from? To get a $7/2^-$ state by simple one step knockout, we must find an $f_{7/2}$ proton in the ground state of ^{12}C. In fact, to couple it to 0^+ we need to find <u>two</u> $f_{7/2}$ particles in the ground state. This is very unlikely given the current understanding of p-shell nuclei. Estimations of the major shell mixing

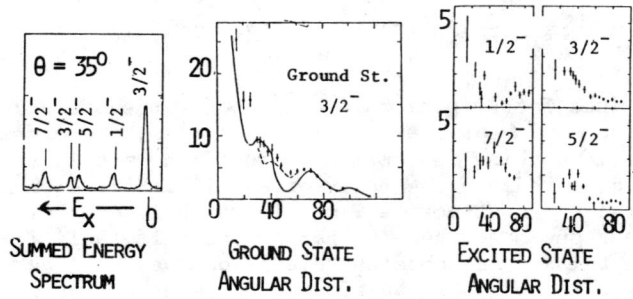

Fig. 5. Coplanar symmetric $^{12}C(p,2p)^{11}B$ at 50 MeV.

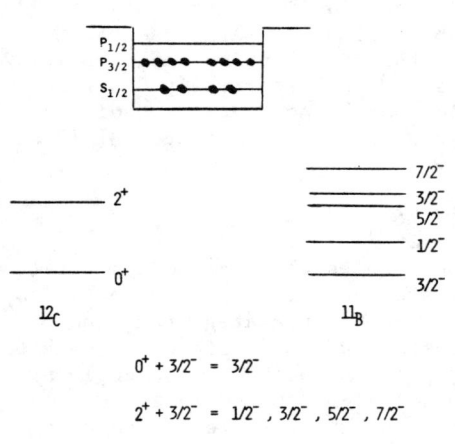

Fig. 6. Structure of the ^{11}B spectrum.

which might be expected to occur indicates that there is far too little to explain the observed results[5].

The simplest way to explain the structure of the spectrum of ^{11}B is to couple a one hole state to the first excited state of ^{12}C. This is shown in Fig. 6. The ground state can be interpreted as a $p_{3/2}$ hole in a ^{12}C closed shell and all of the first four excited states as the coupling of a $p_{3/2}$ hole to the 2^+ excited state of ^{12}C. To reach the excited states, we have to go through two steps: first, the ^{12}C must be excited to its 2^+ state, then the $p_{3/2}$ particle knocked out (or vice versa). This experiment at 50 MeV gives strong evidence that such processes take place.

The two-step process can be expected to show a strong energy dependence. Since cross sections generally fall with energy, the two-step processes should drop with increasing energy faster than the one-step. This is borne out in a comparison of the 50 and 100 MeV data sets for $^{12}C(p,2p)$ as shown in Fig. 7 from ref. 6. Remember at the lower energy, the cross sections to the ground and excited states are comparable. Here we see at the higher energy, the excited states are suppressed by a factor of 10.

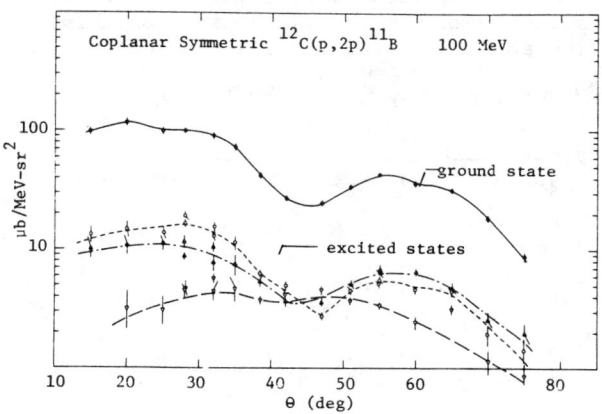

Fig. 7. Angular correlation for ^{12}C (p,2p) ^{11}B at 100 MeV.

This points out one important advantage that knockout reactions have over pickup for studying hole states. In a pickup reaction, two-step processes become relatively _more_ important as energy increases because the direct one step matrix element has an increasingly bad momentum mismatch as the energy goes up. This favors overcoming the mismatch by taking two small steps. At 700 MeV in (p,d) the "two-step" states are still excited as strongly as the "one-step" states.[7] In (p,2p) there is no momentum mismatch in the direct matrix element so two-step processes should become less important at high energies.

A second failure of the DWIA approach arises in the analysis of (p,2p) reactions at energies below 150 MeV. For example, the DWIA fit to ^{4}He(p,2p) from 65 MeV to 600 MeV was shown by Roos yesterday and gives a good qualitative representation of the data.[8] The large discrepancy between the normalization at the low energy end and at the high energy end is well accounted for. When we look

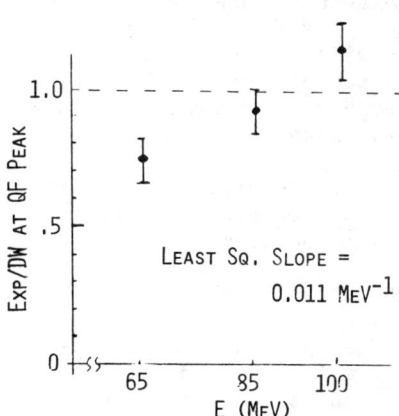

Fig. 8. Energy dependence of spectroscopic factors in ^{4}He (p,2p)^{3}H.

at the low energy end in detail, some discrepancy still remains. This is shown in Fig. 8. We note that there is a change of about 50% in the spectroscopic factor calculated using the DWIA at 65 and 100 MeV. In the case of ^{4}He, there are no low lying excited states which can easily be excited to lead to a two-step process. Three possible reasons for the anomaly are: (1) systematic experimental error, (2) the inadequacy of the approximations made in evaluating the DWIA, or (3) the presence of true three-body reaction mechanisms. An example of this would be breakup followed by modification of the three-body wave functions by interactions not representable as the product of two-body effects. This reaction mechanism would require something more sophisticated than a two-body model for its description.

WHAT CAN HAPPEN: MORE SOPHISTICATED REACTION MECHANISMS

Let's now consider why our reaction mechanism might be failing. Below we summarize our discussion of the nuclear ranges involved.

D_0	= Range of the strong part of the nuclear force		1 fm
D_0'	= Effective size of a nucleon	$(\sigma_{N-N}^{tot}/\pi)^{1/2}$	$\sim$ 1 fm
D_1	= Average separation of nucleons in a nucleus	$(1/\rho_{NM})^{1/3}$	$\sim$ 1.8 fm
D_2	= Mean free path of a nucleon in Nuclear Matter	$1/(\rho_{NM}\sigma_{N-N}^{tot})$	$\sim$ 1.5 - 2.5 fm

$$D_0 = D_0' < D_1 < D_2$$

D_0 is the range of the strong part of the nuclear force as estimated from semi-phenomenological potentials models (with some input from meson theory). D_0' is what we get by taking a typical nucleon-nucleon total nuclear cross section (at any energy from about 100 MeV to 1000 MeV) and setting it equal to πr^2. Fortunately, this is consistent with the first line. D_1 is an estimate of average inter-nucleon spacing in a nucleus at typical densities, and D_2 is the mean free path obtained from the classical formula using the density and cross section mentioned above. For our choice of effective transition operator to be good, we want $D_0 \ll D_1$. For our distorted wave approach to work, we want $D_1 \ll D_2$. Actually D_2 doesn't have to be long, since a scattering leading to an excitation which has little chance of making the transition we want doesn't hurt. It can be easily taken into account using distorted waves. What matters is the mean free path between scatterings which lead to important excitations; ones which can frequently make the transition we're looking for.

Therefore, simply from our qualitative range arguments, we expect there to be possible modifications of our DW picture; first, by modifications of the effective transition operator due to the presence of other particles; and second, whenever there are particular excited states formed strongly which can lead to our final state.

We note that strongly coupled states can prevent a DWIA fit even if strong transitions do not occur from the excited state to the final state we want. In this case, the DWIA is formally correct, but the optical potential required will have a large non-local component which cannot be mocked up by the local potentials universally used.

In the two-step case discussed, let's look at what happens on ^{12}C when a proton comes in. Figure 9 (from ref. 9) shows the angular distributions of the outgoing particles indicated for the strongest discrete final state which has been observed. We see that the two strongest processes are inelastic scattering and pickup of a neutron. To get to a final knockout state after we picked up a neutron, we'd have to have a charge exchange reaction as a second step and this would suppress the contribution. Therefore, we only consider inelastic scattering. In fact, at low energies, the largest inelastic scattering tends to occur to the giant resonances which are above

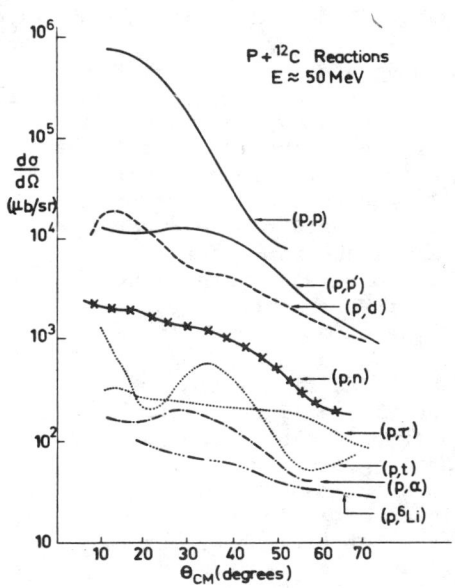

Fig. 9. $p + {}^{12}C$ reactions, $E \approx 50$ MeV.

particle emission threshold and therefore are not shown in the figure. The effects of these inelastic scatterings on knockout have been considered for the $^{16}O(p,2p)^{15}N(g.st.)$ at 45 MeV and for the $^{12}C(p,2p)^{11}B(g.st.)$ at 50 MeV. The effect seems to be non-trivial as seen in Fig. 10 (from ref. 10), but the calculations are extremely difficult and the results inconclusive at this time. The contribution to the excited state knockout, which is the interesting one, has not been considered for this process.

The effect of inelastic scattering through the first 2^+ of ^{12}C followed by knockout is being considered by Walker and Picklesheimer[11] and is found to

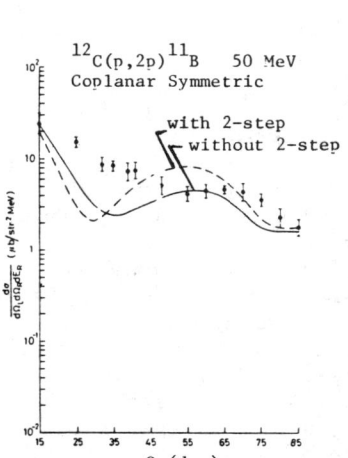

Fig. 10. Effect of two step through giant quadrupole inelastic scattering on $^{12}C(p,2p)^{11}B$.

give cross sections at 50 MeV and 100 MeV which are consistent with the data. These calculations are interesting but are presently in a preliminary stage and cannot be considered conclusive. Further work is needed.

Now let's consider the type of reaction mechanism that could lead to the energy dependence in the ^{4}He (p,2p) and analyzed by Roos. In this case there is a significant absence of strongly excited two cluster channels. We are forced to consider more general reaction mechanisms. As in the previous case, we ask the question: What happens most in the reaction or, more precisely, are there any strong processes occuring which are not two-cluster in character. As an example, let's look at inelastic proton scattering over the full spectrum of possible energy losses. Figure 11 (from ref. 12) shows the energy spectra resulting at various angles when 160 MeV protons are incident on ^{12}C. Scattering with no energy loss (elastic scattering) is at the extreme right with low energy outgoing protons on the left. The qualitative structure of these spectra is shown schematically in Fig. 12 (from ref. 13). We see the elastic scattering on the far right, then a few

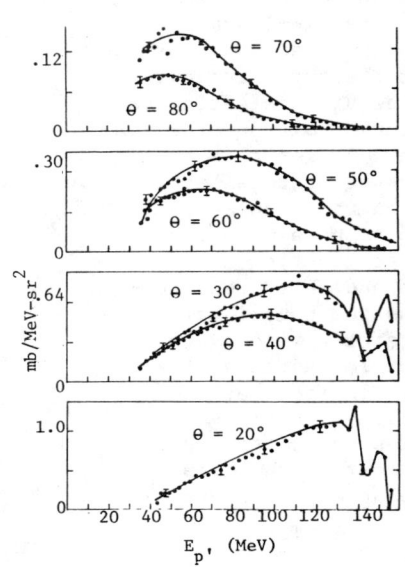

Fig. 11. Proton energy spectra from p + ^{12}C at 160 Mev.

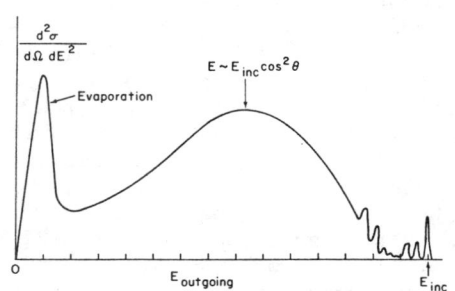

Fig. 12. Schematic structure of proton energy spectra in p + light nucleus scattering.

peaks corresponding to direct reactions to low lying states of the target, then a large peak in the continuum and finally, some low energy boil-off protons from compound nuclear processes. Clearly, there is a large, important reaction taking place. The large peak corresponds to direct quasi-free knockout of a target nuclear. This interpretation is supported by the angular dependence of the peak energy which is shown in Fig. 13. The x's indicate the observed peak energy of the singles spectra with the boxes giving an indication of the peak's width. The solid line gives the energy which would be observed if the process were simply free nucleon-nucleon scattering.

Given the fact that there is a lot of quasi-free knockout taking place we have to rethink our reaction mechanism criteria carefully. We may consider each different division of energy among the three particles as a separate channel. The probability

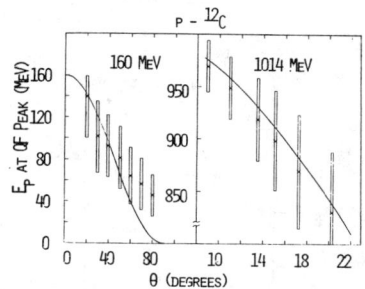

Fig. 13. Peak positions in inelastic proton spectra compared to free p+p kinematics.

is small that each of these channels, once they are reached, will scatter into the specific three-body final state we are looking for. However, there are a lot of them. The sum total of rescatterings into the channel we're observing might be very important. We expect that at high energies, if straight-line semiclassical paths dominate, that each scattering will have as its primary effect an attenuation of the projectile's energy. There should be only a small number of indirect paths bringing us to the desired final state. This suggests that at "high energy" (whatever that means) the DWIA should work. At what energies might these three-body reaction mechanisms be important? Unfortunately my intuition of three-body processes is not well enough developed that I can construct a simple argument for you. Instad, let me show the results of a model study.[14] The model attempts to mock up the ^{4}He (p,2p) experiment as closely as possible while still including three-body rescattering effects to all orders. The model consists of two (identical) spinless protons and an inert triton. Separable potentials of Yamaguchi form are chosen to simplify the calculation. In this model an exact DWIA can be calculated as well as the full solution of the Faddeev equations. The DWIA includes the distortions arising from the p-t interaction averaged over the ^{4}He wave function and the p-t final state interaction. The complete fully-off-shell T matrix is included as are recoil effects. The exact and DWIA results are compared in Fig. 14 at 65 and 100 MeV. The DWIA is too big at low energies and comes down towards the data as energy increases. The rate of change of the "spectroscopic factor" with energy (i.e., S = Exact/DW at the QF peak) is about 0.6%/MeV compared to Roos' 1.1 ± 0.5%/MeV. (See Fig. 8.) The rate of change is of the right order of magnitude and has the correct sign. The result is extremely gratifying considering the simplicity of the model. Just to show that the model is somewhat realistic, the energy sharing distribution predicted is compared with the ^{4}He (p,2p) data in Fig. 15. The shape given by the Faddeev model is quite good, although the normalization is off by a factor of 3.

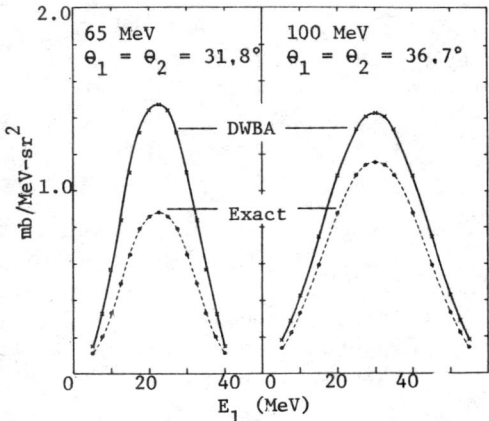

Fig. 14. Energy sharing coincidence spectra for ^{4}He(p,2p)^{3}H in a three-body model.

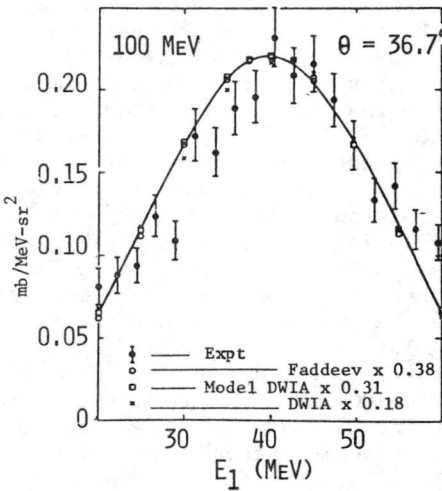

Fig. 15. Energy sharing spectra for ^{4}He(p,2p)^{3}H.

CONCLUSIONS

In summary, the (p,2p) reaction seems to be a very rich one. A lot of things happen together, but in a way which permits them to be disentangled. At energies of 200 MeV or above the factorized DWIA should work well. Roos' calculations of the energy sharing geometry in the factorized approximation combined with Koshel's justification of the factorization are particularly encouraging. This should then permit the new machines at Indiana, TRIUMF, Los Alamos, etc., to do true spectroscopic work and help fulfill the promise the reaction had many years ago.

Once that information is obtained, the reaction can be extended to other energies and other regions of the phase space in order to investigate off-shell effects and reaction mechanisms.

As far as off-shell effects go, let me stress that the determination of the true nucleon-nucleon force is still the most important problem facing nuclear physics today. Without a reliable nuclear force, the transition from microscopic to macroscopic (i.e., from the two-body interaction to the many-body properties) is impossible. The determination of the on-shell amplitude even with perfect accuracy at all energies (which is by no means a forseeable goal) is not sufficient to determine the potential. At this time there is no single experimental determination of an off-energy-shell nucleon-nucleon matrix element. Average properties can be assessed with binding energy calculations but it's very difficult to get anyting clean, both because of the degree of averaging and because of the uncertainties in the many-body theory. The other reactions which have been proposed to study off-shell behavior directly, bremsstrahlung and breakup in the three-nucleon system, have been discredited as off-shell probes in the past few years.[15,16] If we can handle the reaction mechanism in (p,2p) the off-shell nature of the reaction could be its most valuable aspect.

Finally, the three-body nature of the reaction which I have speculated may be appearing at energies below 100 MeV may permit us to learn something about how nucleons behave at low energies and the extent to which the picture of the independently correlated pair,[17] which has dominated thinking in nuclear physics for so long, actually holds in real nuclei.

REFERENCES

1. Reproduced from *Nuclear Structure* by Aage Bohr and Ben R. Mottelson with permission of Addison-Wesley/W.A. Benjamin, Inc. Advanced Book Program, Reading, Mass.
2. E. F. Redish, G. J. Stephenson, and G. M. Lerner, Phys. Rev. $\underline{C2}$, 1665 (1970).
3. E. F. Redish, Phys. Rev. Lett. $\underline{31}$, 617 (1973).
4. H. Pugh, D. Hendrie, M. Chabre, E. Boschitz, and I. McCarthy, Phys. Rev. $\underline{155}$, 1054 (1967).
5. G. J. Stephenson, Jr., *Int. Nucl. Phys. Conference*, Gatlinburg, Tenn., R. L. Becker, ed. (Academic Press, N.Y., 1962).
6. R. K. Bhowmik, et al., Nucl. Phys. $\underline{A226}$, 365 (1974).

7. S. D. Baker, et al., Phys. Lett. $\underline{52B}$, 57 (1974).
8. P. G. Roos, Phys. Rev. $\underline{C9}$, 2437 (1974).
9. E. F. Redish, Nucl. Phys. $\underline{A235}$, 82 (1974).
10. P. Wright, Ph.D. Thesis, Flinders University, 1975, unpublished.
11. G. Walker, private communication.
12. N. S. Wall and P. G. Roos, Phys. Rev. $\underline{150}$, 811 (1966).
13. N. S. Wall, Univ. of Maryland tech. rpt. no. 76-060, 1975, unpublished.
14. S. K. Young and E. F. Redish, Phys. Rev. $\underline{C10}$, 498 (1973).
15. L. Heller, in <u>The Two-Body Force in Nuclei</u>, S. M. Austin and G. M. Crawley, eds. (Plenum, N.Y., 1972).
16. D. D. Brayshaw, Phys. Rev. $\underline{C13}$, 1835 (1976).
17. H. Bethe and J. Goldstone, Proc. Roy. Soc. $\underline{A238}$, 551 (1957).

QUASI-FREE (p, pα) SCATTERING AT 157 MeV
P. Radvanyi,
Institut de Physique Nucléaire, Orsay, France.

This is a short summary of results obtained in experiments performed with the Orsay Synchrocyclotron on the (p, pα) reaction on several even-even nuclei.[1]

Energy spectra have been measured around the quasi-free kinematic conditions for the reactions ^{24}Mg(p, pα)^{20}Ne, ^{28}Si(p, pα)^{24}Mg, ^{40}Ca(p, pα)^{36}Ar and ^{58}Ni(p, pα)^{54}Fe. Recoil momentum distributions have been obtained for the 0^+ ground state and the 2^+ first excited state of the three s-d shell nuclei.

The reaction appears to have a strong quasi-free behaviour. Using a distorted wave analysis with the Maryland Chant code[2], we obtained the relative spectroscopic factors given in Table I, the ^{28}Si → ^{24}Mg (0^+) +α transition being taken as unity; these relative spectroscopic factors have been compared with SU(3) model predictions:

TABLE I

RELATIVE $S_α$	0^+ states			2^+ states		
	^{24}Mg	^{28}Si	^{40}Ca	^{24}Mg	^{28}Si	^{40}Ca
Experiment DWIA	$1.0^{+}_{-}0.4$	1	$2.1^{+0.6}_{-0.5}$	$0.9^{+0.5}_{-0.5}$	$1.5^{+0.6}_{-0.4}$	$3.8^{+2.2}_{-1.8}$
SU(3)	0.9	1	1.0	0.1	1.2	5.0

There appears however a disagreement in shape between experiment and the DWIA quasi-free calculations at low recoil momentum for the 2^+ states. This might result from another reaction mechanism superposed on ordinary quasi-free scattering: a 2 step mechanism proceeding through knockout preceded or followed by an inelastic scattering, or a quasi-free scattering on 2 protons and 2 neutrons with J =2, L = 0.

REFERENCES

1. D. Bachelier, J. L. Boyard, T. Hennino, H. D. Holmgren, J. C. Jourdain, P. Radvanyi, P. G. Roos, M. Roy-Stephan, Nucl. Phys. A268 (1976)488.
2. N. S. Chant, P. G. Roos, Proc. 2nd Int. Conf. on clustering in nuclei, Univ. of Maryland (1975)p. 265.

GREEN'S FUNCTION APPROACH TO THE $(\hbar\omega, 2e)$-REACTIONS*

Peter Winkler
Universität Erlangen-Nürnberg, West Germany

INTRODUCTION

According to the NSF proposal for this workshop the investigation of (e,2e)-reactions is still in its infancy. Using the same time scale one might say that double photoionization (abbreviated henceforth (γ,2e) is still an unborn child. Why then talk about it now and here? One obvious reason is the two-electron continuum as the final state of both reactions. The fact that (e,3e)-reactions under photon simulating dynamical conditions are being set up by groups who also investigate (e,2e)-reactions is another example of the close relationship. The main emphasis of this report, however, is the recognition that the correlated motion of electrons is described only incompletely by the distribution of the one-electron momentum alone. The distribution of relative momentum of the electrons in the system is another important piece of information which has to supplement the former. We are further convinced that any kind of collective electron motion will have its strongest and most direct impact upon this latter quantity. Collectivity is certainly less sensational in atomic and molecular physics than it has proven to be in nuclear and solid state problems. Nevertheless, there is at least the whole field of transport phenomena within large molecules open to investigation from this point of view. It is still to be determined whether (γ,2e)-experiments are capable of supplying the desired pieces of information. From the little which is known so far experimentally and theoretically a thorough judgement is not yet feasible. One can, however, make some statements concerning the required degree of sophistication of the analysis of (γ,2e)-experiments if more than an explanation of gross features of experimental data is aimed at.

2. Short review of early work on the (γ,2e)-reactions

First experimental evidence for the importance of e-e-correlation in (γ,2e)-reactions is given by Carlson's measurements[1] of the relative abundances of doubly to singly charged ions of the rare-gas atoms He, Ne, and Ar. The photon energy ranges from 3 to 8 times the threshold energy for double photoionization. In all three cases the findings are such that they cannot be explained by electron shake-off theory on the basis of an independent particle model. (Shake-off theory reflects the physical situation of the electron orbitals rearranging in the changed average field after one electron has been emitted.)

*Work performed under the auspices of and sponsored by the Deutsche Forschungsgemeinschaft

Carlson's results for the He atom have been explained by Byron and Joachain[2] who include correlation on both the initial and to some extent in the final state (a six-term Hylleraas-type wave function for the former whereas a θ_{12}-dependent term is included in the latter). The importance of a correlated analysis can be seen from fig. 1.

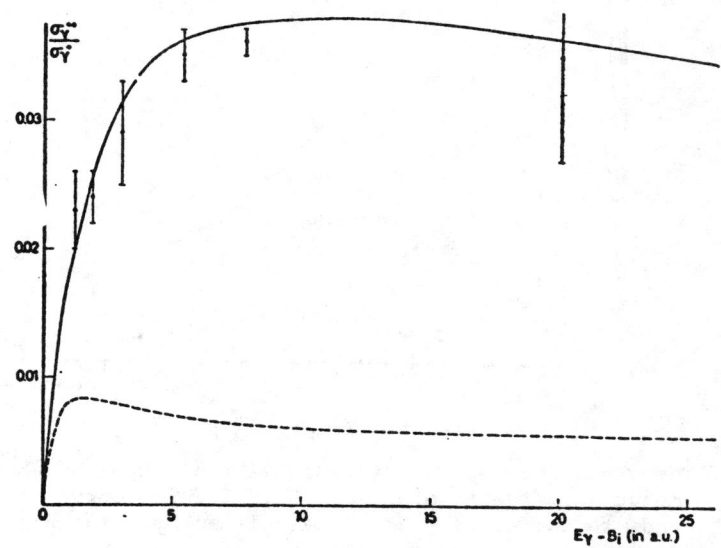

Fig. 1, taken from Ref. 2

The ratio $\rho_\gamma = \sigma_\gamma^{++}/\sigma_\gamma^{+}$ as a function of $E_\gamma - B_i$. The solid curve represents the theoretical calculation of this paper, using a fully correlated initial-state wave function. The dashed curve corresponds to a Hartree-Fock ground-state wave function. The points are the experimental results of Ref. 1.

Two further observations of Byron and Joachain seem to reflect general features and are worth mentioning here: a) The discrepancy between cross-sections calculated from dipole matrix elements in length and momentum form. The size of the deviation is remarkable since one feels that the wave functions employed are fairly accurate. b) The inclusion of angular correlation in the final state has a strong influence on the form of the position results but changes the form of the momentum cross sections only a little. The conclusion that the momentum form should be given preference is debatable and certainly not generally true. Further improvement of the ground state wave function does not improve the results remarkably as shown by Brown.[3] Meanshile a considerable amount of experimental effort

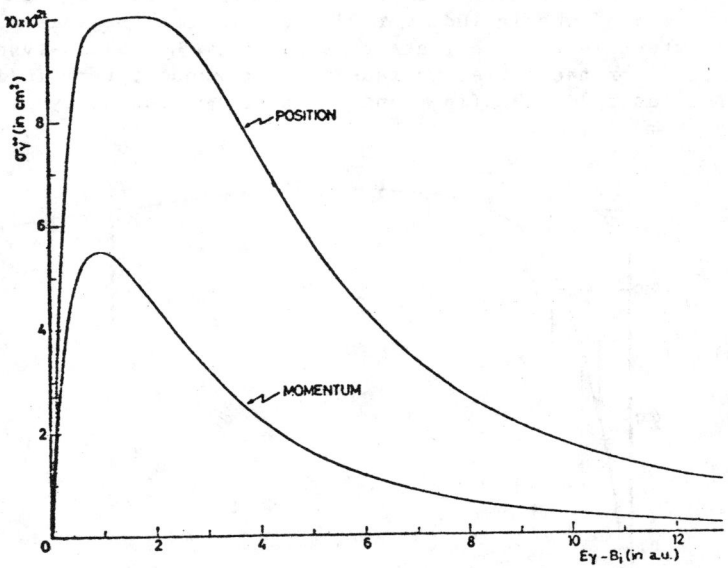

Fig. 2, taken from Ref. 2.

The cross section σ_γ^{++} for double ionization by photon impact as a function of $E_\gamma - B_i$. The top curve corresponds to the cross section calculated by using position dipole matrix elements. The bottom curve shows the cross section obtained from momentum matrix elements.

has been devoted to the investigation of multiple ionization phenomena in the outer shell of the rare gas atoms[4,18,19,23,24]. A theoretical interpretation of such data has been carried out only for the Ne atom by Chang and Poe.[5] Let us have a closer look at this beautiful and detailed analysis.

3. The analysis of Chang and Poe

Within the framework of a low order many-body perturbation theory (MBPT) the relative importance of different contributions to the (γ,2e) cross section for Ne is investigated for a range of photon energies (from threshold up to 220 eV above). There are four predominant contributions each of which is represented by a typical diagram in the following figure:

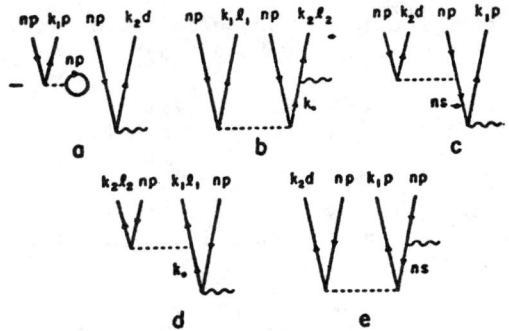

Fig. 3, taken from Ref. 5

Diagrams representing important contributing effects in double photoionization of rare-gas atoms. The wavy lines stand for photon interaction H_γ; dashed lines represent the electron-electron interaction.

a) <u>Core rearrangment</u>: This diagram represents the shake-off process of a second electron due to a change in screening. This mechanism is a major contribution to the (γ,2e) cross section in all energy ranges but it cannot describe the observed cross sections alone. (We realize that it is due to the <u>change</u> of the SC-field that a closed p-hole-loop appears explicitly in a diagram which in addition is unlinked.) b) <u>Ground state correlation</u>: Diagrams b and e represent electron correlation in the initial state. In b two valence shell electrons interact via r_{12}^{-1}. The incoming photon supplies the momentum and energy required for both electrons being put on energy shell. In e the physical situation is similar. The basic difference is that we start with two electrons from different subshells scattering with each other which is a much less probable process. Diagram e is negligible here. c) <u>Virtual Auger transitions</u>: One s electron is photoionized. One p electron drops into the s hole, another p electron is emitted. The difference to the Auger effect is that in the intermediate state there is no energy conservation.
d) <u>Inelastic internal collisions</u>: This is a direct collision of the outgoing with one of the atomic electrons. It is very probable at low energies and becomes negligible only at about 120 eV above threshold.

Each of these diagrams represents a transition $t_\alpha^f(k_1 k_2)$ to a particular (but always the same) final state f. The sum of them all makes up an element of the transition matrix

$$T_f(k_1 k_2) = \sum_\alpha t_\alpha^f(k_1 k_2)$$

Various cross sections are calculable from this quantity: a) The total cross section for double photoionization at a given photon energy E_γ (dipole momentum approximation)

$$\sigma(E_\gamma) = \frac{16\alpha\, a_0 e^2}{E_\gamma} \sum_f \int \frac{dk_1}{k_2} |T_f(k_1 k_2)|^2$$

b) The cross section for two photoelectrons at a specified combination of the two asymptotic energies $k_1^2/2$ and $k_2^2/2$ (with E_γ given and fixed):

$$\sigma(k_1 k_2) = \frac{16\alpha\, a_0 e^2}{E_\gamma} \sum_f \frac{1}{k_1 k_2} |T_f(k_1 k_2)|^2$$

c) The photoelectron energy spectrum at given E_γ:

$$\frac{d\sigma}{d\epsilon}(E_\gamma) = \sigma(k_1 k_2) + \sigma(k_2 k_1)$$

This is displayed in the next figure for various photon energies.

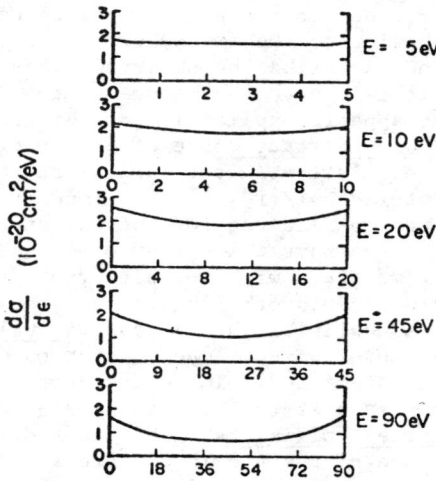

Fig. 4, taken from Ref. 5.

Calculated photoelectron spectra for total kinetic energy $E = 5, 10, 20, 45,$ and 90 eV.

The investigation of the energy dependence of the individual contributions is the key information of this analysis since its gross features will very probably carry over to other systems. It is presented in the following graph:

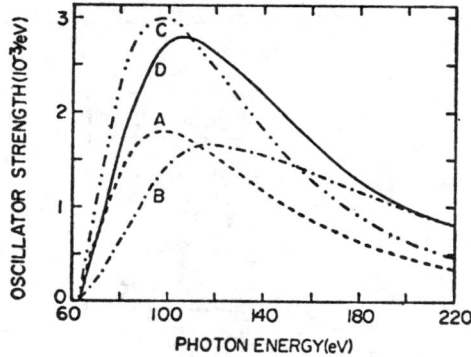

Fig. 5, taken from Ref. 5.

Calculated oscillator strengths: curve A——— core rearrangement [diagram (a)]; curve B——— core rearrangement, ground-state correlations, and virtual Auger transition [diagrams (a) + (b) + (c)]; curve C——— core rearrangement, virtual Auger transition, and inelastic internal collision [diagrams (a) + (c) + (d)]; curve D——— total contribution [diagrams (a) + (b) + (c) + (d) + (e)].

Our main concern is the investigation of the problem how to use double photoionization as a probe for ground state correlation. From fig. 4 we can easily find out by comparing curves B and D that the region below $E_\gamma \lesssim 90$ eV (above threshold) is strongly effected by contributions due to inelastic internal collisions which are negligible at energies above 120 eV (above threshold). The core rearrangement (curve A) is nowhere really small whereas the virtual Auger contributions constitute only a minor correction at high energies as can be seen from the difference of curves C and A.

We conclude that investigations of ground state properties require energies $E_\gamma \gtrsim 100$ eV. If on the other hand studies of the reaction mechanism were our declared aim, we would preferentially turn to low energy situations. These qualitative features are not really new or unexpected. The exact evaluation by Chang and Poe, however, is new and extremely helpful for our understanding of the $(\gamma, 2e)$-reaction at corresponding energies.

The comparison of Chang and Poe's findings with experimental data is presented in the following figure.

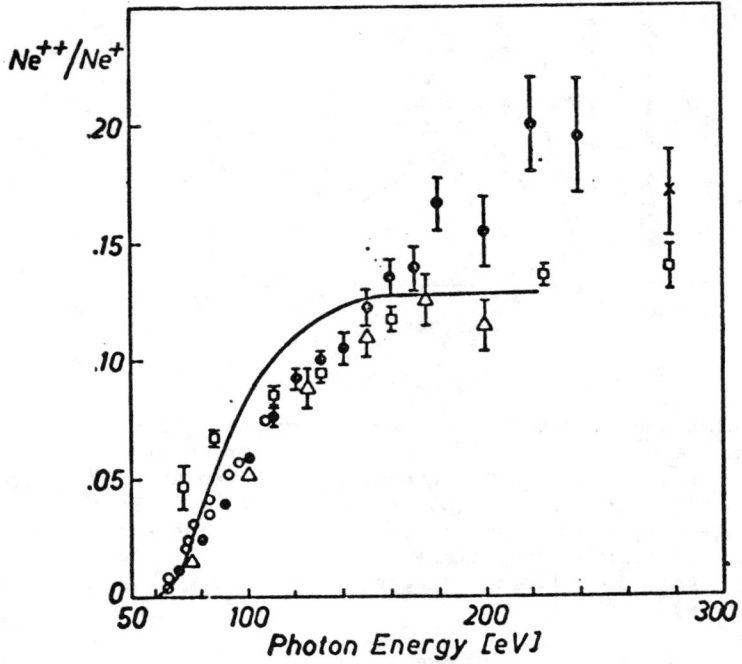

Comparison of Chang and Poe's results (solid line) with experimental data.

(Figure supplied by V. Schmidt)

□ Carlson, Ref. 1;
△ Schmidt et al., Ref. 19;
● Wight and Van der Wiel, Ref. 18;
○ Samson and Haddad, Ref. 4;
× Lightner et al., Ref. 4.

The agreement is fairly good although not yet completely satisfactory. Possible explanations for the deviations of Carlson's data are discussed by the authors.

4. A discussion of this analysis

A few remarks have to be made concerning the conclusiveness of Chang and Poe's study as well as the corresponding experimental data. We face the fact that a low order calculation is apparently in fairly good agreement with what is expected to be the outcome of a rather complicated physical process. If this happened to be the final answer we would conclude that the sensitivity of the experiment does not allow for the study of details of the e-e interaction in the system. More selective experiments are required. They will have to go hand in hand with careful theoretical analysis. For example, we know that core-polarization effects are present in the potential felt by one electron in the field of the core as well as in the interaction two electrons experience in the presence of the ionic core. The former processes have been studied by Chang and Poe[6] in a many-body theory of single photoionization of Li and found to be non-negligible. It is only through the inclusion of those and higher order short range correlations that approximate agreement of about 1.5% in the length and momentum forms of the total ionization cross sections is achieved. Taking into account that in the Li case the remaining ion is in a $1s^2$ configuration whereas in the Ne atom the two electrons leave an open shell system which obviously is more easily polarized we may conclude that these contributions will have to be included in a more thorough (and much more difficult!) treatment. As to the experimental situation, we feel that measurements of the angular correlation of the two outgoing electrons have to come next. It is very likely that in the measured total cross section cancellation of competing contributions occurs and thus this experiment in principle does not allow for the detection of the angular dependence of the effective interaction. Although Ref. 5 does not give information about possible discrepancies between the length and momentum results it can be surmised from the earlier work of Byron and Joachain on He and from a very recent calculation of Carter and Kelly[7] on the carbon atom that these deviations are on the order of 50 per cent.

In the following sections the ideas of an improved treatment are outlined for systems with two electrons outside of a closed shell. It turns out that for those systems the problem can be formulated in a way which allows for an approximate inclusion of higher order contributions as discussed above without encountering considerably more numerical difficulties than in Ref. 5. These calculations are presently carried out in Erlangen for Be and Li⁻. In order to make visible how the present approximation is embedded in a general manybody Green's function approach (not necessarily MBPT) a few general facts about connection of Green's functions and effective potentials are recapitulated.

5. Green's function and effective potentials

In the following notation we stress the character of the one-electron Green's function as a probability amplitude for the propagation of an additional electron (or: hole) from space-time-spin point 1' to 1 (or: 1 to 1') under full interaction with the underlying system:

$$G(1,1') \equiv i^{-1}\theta(t-t')\langle\Phi_0^N|\psi(\underline{r})e^{-i\hat{H}(t-t')}\psi^+(\underline{r}')|\Phi_0^N\rangle e^{iE_0^N(t-t')}$$

$$- i^{-1}\theta(t'-t)\langle\Phi_0^N|\psi^+(\underline{r}')e^{-i\hat{H}(t'-t)}\psi(\underline{r})|\Phi_0^N\rangle e^{iE_0^N(t'-t)}$$

Here Φ_0^N represents the exact ground state of an atom (or molecule) with N electrons. $\psi^+(\underline{r})$ creates an additional electron at location $\underline{r}$. (ψ is the corresponding destruction operator).

Since the equation of motion for ψ (and therefore also for G) is nonlinear it turns out that the one-particle Green's function is coupled to the two-particle Green's function G_{II} (in general is G_n coupled to G_{n-1} and G_{n+1}). We cannot separate these equations. We can, however, replace the effect of all the higher order Green's functions by an effective potential function Σ of only two (spin-space-time) variables in the sense that if we knew Σ <u>exactly</u> we could solve the following equation for an <u>exact</u> G:

$$[i\frac{\partial}{\partial t} - h(1)]G(1,1') - \int d3\ \Sigma\ (1,3)G(3,1') = \delta(1-1')$$

Any approximation to Σ will yield an approximate G. (In the special case of a local and time-independent potential function we obtain the usual one-particle Schroedinger equation, with h (1) representing the one-electron operators except Σ). Similarly, in the case of the two-particle Green's function G_{II} we obtain the following equation due to Bethe and Salpeter:

$$G_{II}(1,2;1',2') = G(1,1')G(2,2') - G(1,2')G(2,1')$$

$$+ \int d3d3'd4d4'\ G_{II}(1,2;3',4')W(3',4';3,4)G(3,1')G(4,2')$$

which looks much more transparent in a graphical representation:

Again: if we knew the interaction (or: vertex) function W <u>exactly</u> we could solve for an <u>exact</u> G_{II} (using the exact G obtained in the first step).

A variety of approaches to obtain approximations to the key quantities Σ and W can be and have been given in the literature. The decision as to which one is the best depends critically on the investigated problem. For the purpose of using the (γ,2e) reaction as a probe for the pair correlation in atomic or molecular systems a more detailed analysis of Σ and W is required than is incoporated in Chang and Poe's work. In particular polarization effects have to be included. For an extension of Chang and Poe's work on Ne this means the evaluation of a great number of higher order diagrams where both the fermion lines have to be modified by the following type of insertions

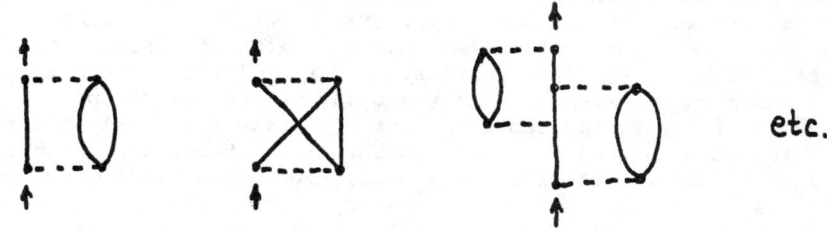

(the arrows indicate where outer fermion lines are to be attached)

and the interaction lines themselves have to be redefined to contain also polarization bubbles

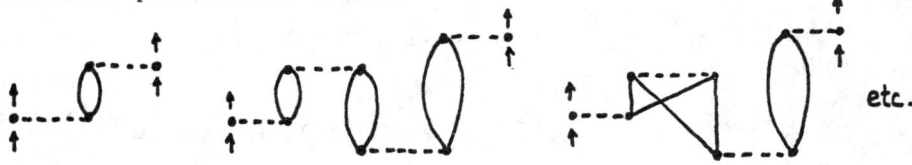

Some of these higher order contributions can be treated with moderate computational effort others cannot. One prominent member of the latter type is the following diagram which corresponds to the physical situation of electron A polarizing the core. This change of the electron distribution is experienced by electron B some time later. In the meantime, however, both electrons interact through the Coulomb potential:

6. Connection with pseudo potential methods

There is no doubt that the analysis of higher-order energy dependent contributions has to be and actually is being carried through for a refined interpretation of detailed experimental data expected to come. For atoms with two valence electrons outside a closed shell, however, a reasonable approach is to "dress" the fermion and interaction lines for the valence electrons by redefining them through semiempirical or pseudo potentials which simulate the effect of the polarizable inner shells. Potentials of this type have been successfully used before, mostly for bound state investigations. The usual argument of the characteristic velocities of the core electrons being large compared to the valence electrons such that the time dependent (or equivalently: energy dependent) interaction may well be simulated by a static potential will hold for the (γ,2e) reaction to the same extent as it holds for the investigation of bound state properties. A particularly interesting approach has been reported by Csanak and Taylor[8] who, starting from the random phase approximation (RPA) for Σ, derive rigorously the analytical form of a semi-empirical potential which contains practically all the features of previously used pseudo potentials and clarifies in a systematic way all necessary approximations in reducing the non-local energy dependent Σ^{RPA} to a local and static potential. By this derivation from a more comprehensive many-body theory part of the arbitrariness of earlier models is removed. Here we compare the results of Ref. 8 to the model of Laughlin and Victor (Ref. 9):

$$\sum_{Cs\&T}(r) = -\frac{Z}{r} + \sum^{Core} -\frac{1}{4\pi}(\frac{1}{16r_0^3})^2 \mu e^{-2\bar{r}}[P^{(1)}(\bar{r})]^2$$

$$-\frac{\alpha_{dip}}{2r^4}[1-\frac{1}{8}e^{-\bar{r}}P^{(3)}(\bar{r})]^2 - \frac{\alpha_{quad}}{2r^6}[1-\frac{1}{144}e^{-\bar{r}}P^{(5)}(\bar{r})]^2 + \frac{3\beta}{r^6}[1-\frac{1}{8}e^{-\bar{r}}P^{(3)}(\bar{r})]^2$$

$$\sum_{L\&V}(r) = -\frac{Z}{r} + \sum^{Core}$$

$$-\frac{\alpha_{dip}}{2r^4}W_6(\bar{r}) - \frac{\alpha_{quad}-6\beta}{2r^6}W_8(\bar{r}) + U(r|a)$$

Here $\quad W_n(\bar{r}) \equiv 1 - e^{-\bar{r}^n}$

$\bar{r} \equiv \frac{r}{r_0}$ (not necessarily the same parameter r_0 in both models!)

$$U(r|a) \equiv (a_0 + a_1 r + a_2 r^2)e^{-kr} + (a_0' + a_1' r)e^{-k'r}$$

$P^{(\ell)}(\bar{r})$ polynomial of ℓ-th degree in $\bar{r}$ (a corrected version is given in Ref. 10)

The quantities α_{dip}, α_{quad} and μ are the static dipole polarizability, the quadrupole polarizability and the monopole hyperpolarizability, respectively. The last term of the first equation is the first nonadiabatic correction with β parameterizing the dipole oscillator strength and the energy of core dipole transitions.

In Laughlin and Victor's approach the non-adiabaticity is in some sense taken into account by readjusting the quadrupole term through the parameter β. We notice explicitly that the arbitrary cut-off functions W_n, which have to be introduced in order to avoid divergencies near the origin are <u>derived</u> in Csanak and Taylor's approach. The potential correction term $U(r|a)$ is used in Ref. 9 to optimally fit several low lying eigenvalues of each symmetry to the observed Be^+ and Mg^+ spectra. Csanak's analysis has recently been extended to cover the derivation of a semi-empirical interaction potential. (Wang et al., Ref. 10). Again we compare their result to Laughlin and Victor's form:

$$W_{Wang} = \frac{1}{r_{12}} - \frac{1}{4\pi}(\frac{1}{16 r_0^3})^2 \mu e^{-\bar{r}_1} e^{-\bar{r}_2} P^{(1)}(\bar{r}_1) P^{(1)}(\bar{r}_2)$$

$$- \frac{\alpha_{dip}}{r_1^2 r_2^2}[1 - \frac{1}{8} e^{-\bar{r}_1} P^{(3)}(\bar{r}_1)][1 - \frac{1}{8} e^{-\bar{r}_2} P^{(3)}(\bar{r}_2)] P_1(\hat{r}_1 \cdot \hat{r}_2)$$

$$- \frac{\alpha_{quad}}{r_1^3 r_2^3}[1 - \frac{1}{144} e^{-\bar{r}_1} P^{(5)}(\bar{r}_1)][1 - \frac{1}{144} e^{-\bar{r}_2} P^{(5)}(\bar{r}_2)] P_2(\hat{r}_1 \cdot \hat{r}_2)$$

$$+ 3\beta[\frac{1}{r_1^2 r_2^4} + \frac{1}{r_1^4 r_2^2}][1 - \frac{1}{8} e^{-\bar{r}_1} P^{(3)}(\bar{r}_1)][1 - \frac{1}{8} e^{-\bar{r}_2} P^{(3)}(\bar{r}_2)] P_1(\hat{r}_1 \hat{r}_2)$$

$$W_{L\&V} = \frac{1}{r_{12}}$$

$$- \frac{\alpha_{dip}}{r_1^2 r_2^2} W_3(\bar{r}_1) W_3(\bar{r}_2) \cdot P_1(\hat{r}_1 \cdot \hat{r}_2)$$

$$- \frac{\alpha_{quad} - 6\beta}{r_1^3 r_2^3} W_4(\bar{r}_1) \cdot W_4(\bar{r}_2) \cdot P_2(\hat{r}_1 \cdot \hat{r}_2)$$

It is important that no new parameters have been introduced. The idea of this approach is to fit the effective potential to experimental one-electron data of Be^+ and Li and then use the so determined parameters to predict typical two-electron properties such as double photoionization cross sections as is done here or energies of two-particle excited (e.g. autoionizing) states as in Ref. 9.

7. The effective dipole operator

It is well known that consistent use of the theory outlined above requires also a modification of the one-particle operators.[29] For the dipole operator a thorough derivation is given in Refs. 22 and 30. The guiding idea in this analysis is the requirement of numerical equivalence of N-particle (i.e. wave function) treatment and the one-particle picture (i.e. Green's function approach) when it is assumed that both results are obtained without approximations:

$$\langle \Phi_n(1,2,\ldots N) | \sum_{i=1}^{N} \hat{r}_i | \Phi_0(1,2,\ldots N) \rangle$$

$$= \int d\underline{r} f_n^*(\underline{r}) \hat{r}_{eff} f_0(\underline{r})$$

Here $f_n(r)$ is the spatial part of the Feynman-Dyson amplitude

$$f_n(\underline{r},t) \equiv \langle \Phi_n^N | \psi(\underline{r},t) | \Phi_0^N \rangle$$

The explicit form of the effective dipole operator given in Ref. 9 and used for the Be and Li$^-$ calculations is

$$\hat{r}_{eff} = \hat{r}\{1 - \frac{\alpha_d}{r^3}[1 - \exp\{-(\frac{r}{r_0})^3\}]\}$$

In lowest order the use of this effective operator may be visualized as an approximate inclusion of the following diagrams

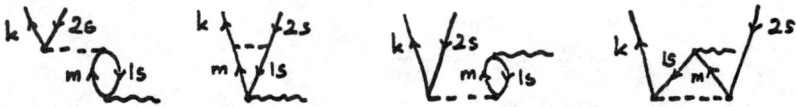

instead of only

Here the diagrams of the Be (or Li$^-$) problem are given.

8. The effective interaction and the pair correlation function

Why this interest in the effective interaction? To answer this question let us first have a look upon some familiar relations from the theory of a homogeneous electron gas. The pair correlation function $g(\underline{r}_{12})$ is obtained by Fourier transforming the energy averaged form factor

$$S(q) = \frac{1}{N} \int_{-\infty}^{\infty} S(q,\omega) d\omega$$

$$\tilde{S}(\underline{r}_{12}) = \delta(\underline{r}_{12}) + (N-1)g(\underline{r}_{12})$$

Here $S(q,\omega)$ is the dynamical form factor. N is the number of electrons per unit volume. Although some information is lost in going from the dynamical to the static form factor this step is consistent with our approximation and even the time independent pair correlation function is a very important quantity in the sense that a number of physical properties of the system can be calculated once g is known, e.g. its ground state energy:

$$E_{g.s.} = N\langle T\rangle + \frac{1}{2}N(N-1)\int W(\underline{r}_{12})g(\underline{r}_{12})d\underline{r}_{12}$$

Here $W(\underline{r}_{12})$ is the elementary interaction, i.e. the Coulomb interaction if the system is an electron gas. We are now going to replace $g(\underline{r}_{12})$ by its Hartree-Fock approximation $g^0(\underline{r}_{12})$ in such a way that

$$\int W(\underline{r}_{12})g(\underline{r}_{12})\,d\underline{r}_{12} = \int W^{eff}(\underline{r}_{12})g^0(\underline{r}_{12})d\underline{r}_{12}$$

Atoms and molecules, however, have a closer resemblance to an inhomogeneous electron gas. Especially for systems which are too big to allow for a wave function description the density functional approach of Hohenberg, Kohn and Sham (Ref. 12 and 13) seems to be a reasonable starting point for further investigation. In this method the ground state energy is given by

$$E_{g.s.} = E[n(\underline{r})] = \int v(\underline{r})n(\underline{r})d\underline{r} + F[n(\underline{r})]$$

Here $F[n(\underline{r})]$ is a universal functional of the density $n(\underline{r})$ for all Coulombic systems independent of the particular potential function $v(\underline{r})$ which creates the inhomogeneity. It consists of the kinetic energy and the Coulombic interaction energy, which here is split up into the classical part of the exchange and correlation part

$$F[n] = \frac{1}{2}\int \nabla_{\underline{r}} \cdot \nabla_{\underline{r}'} n_1(\underline{r},\underline{r}')\Big|_{\underline{r}=\underline{r}'} + \frac{1}{2}\int \frac{n(\underline{r})n(\underline{r}')}{|\underline{r}-\underline{r}'|} d\underline{r}d\underline{r}'$$

$$+ \frac{1}{2}\int \frac{C_2(\underline{r},\underline{r}')}{|\underline{r}-\underline{r}'|} d\underline{r}d\underline{r}'$$

The quantity $C_2(\underline{r},\underline{r}')$ is the pair correlation function defined in terms of the reduced two- and one-particle density matrices ("2- and 1-matrices").

$$C_2(\underline{r},\underline{r}') = n_2(\underline{r},\underline{r}';\underline{r},\underline{r}') - n_1(\underline{r},\underline{r})n_1(\underline{r}',\underline{r}')$$

Upon coordinate transformation

$$\underline{r}_{12} = \frac{1}{2}(\underline{r}-\underline{r}')$$

$$\underline{R} = \frac{1}{2}(\underline{r}+\underline{r}')$$

$C_2(\underline{r},\underline{r}')$ becomes the new function $g(\underline{r}_{12},\underline{R})$ which local version of the pair correlation function replaces $g(\underline{r}_{12})$ in inhomogeneous systems.

$$\int \frac{C_2(\underline{r},\underline{r}')}{|\underline{r}-\underline{r}'|} d\underline{r}d\underline{r}' = \int \frac{\int g(\underline{r}_{12},\underline{R})d\underline{R}}{r_{12}} d\underline{r}_{12} = \int W(\underline{r}_{12})\bar{g}(\underline{r}_{12})d\underline{r}_{12}$$

We require now

$$\int\{\int W(\underline{r}_{12})g(\underline{r}_{12},\underline{R})d\underline{R}\}d\underline{r}_{12} = \int\{\int W^{eff}(\underline{r}_{12},\underline{R})g^{\circ}(\underline{r}_{12},\underline{R})d\underline{R}\}d\underline{r}_{12}$$

where $g^{\circ}(\underline{r}_{12},\underline{R})$ corresponds to the independent particle approximation n_2^0 of the 2-matrix element

$$n_2^0(\underline{r},\underline{r}';\underline{r},\underline{r}') - n_1(\underline{r},\underline{r})n_1(\underline{r}',\underline{r}') \equiv n_1(\underline{r},\underline{r}')n_1(\underline{r}'\underline{r})$$

In the general case of an inhomogeneous electron gas the effective interaction will be a local function. The quantity n_2^0 is completely determined once we know the momentum wave function. With this concluding remark we want to make contact with the top subject of this workshop after having outlined how the pair correlation function (i.e. the Fourier transform of the distribution of relative momentum) supplements our information upon the system.

REFERENCES

1. T. A. Carlson, Phys. Rev. 156, 142 (1967).
   T. A. Carlson, W. E. Hunt and M. O. Krause, Phys. Rev. 151, 41 (1966).
2. F. W. Byron, Jr., and C. J. Joachain, Phys. Rev. 164, 1 (1967).
3. R. L. Brown, Phys. Rev. A1, 586 (1970).
4. M. O. Krause and T. A. Carlson, Phys. Rev. 158, 18 (1967).
   G. S. Lightner, R. J. van Brunt and D. Whitehead, Phys. Rev. A4, 602 (1971).
   J. A. R. Samson and G. N. Haddad, Phys. Letters 33, 875 (1974).
   M. J. Van der Wiel and G. Wiebes, Physica 53, 225 and 411 (1971).
   F. Wuilleumier and M. O. Krause, Phys. Rev. A10, 242 (1974).
5. T. N. Chang and R. T. Poe, Phys. Rev. A12, 1432 (1975).
6. T. N. Chang and R. T. Poe, Phys. Rev. A11, 191 (1975).
7. S. L. Carter and H. P. Kelly, J. Phys. B: Atom. Molec. Phys. 9, 1887 (1976).
8. Gy. Csanak and H. S. Taylor, Phys. Rev. A6, 191 (1975).
9. C. Laughlin and G. A. Victor, "Atomic Physics", Vol. 3, 247 (1973).
10. S. W. Wang, H. S. Taylor and R. Yaris, Chem. Phys. 14, 53 (1976).
11. W. C. Lineberger, private communication.
12. P. Hohenberg and W. Kohn, Phys. Rev. 136, B864 (1964).
13. W. Kohn and L. J. Sham, Phys. Rev. 137, A1697 (1965) and Phys. Rev. 140, A1133 (1965).

14. B. Brehm and A. Bucher, Int. J. Mass. Spec. Ion Phys. 15, 463 (1974).
    B. Brehm and K. Höfler, Int. J. Mass. Spec. Ion Phys. 17, 371 (1975).
15. H. Hotop and D. Mahr, J. Phys. B: Atom. Molec. Phys. 8, L 301 (1975).
16. J. E. Hansen, J. Phys. B: Atom. Molec. Phys. 8, L 403 (1975).
17. U. Fano, Comments on Atomic and Molecular Phys. Vol. 4, 119 (1973).
18. G. R. Wight and M. J. Van der Wiel, J. Phys. B9, 1319 (1976).
19. V. Schmidt, N. Sandner, H. Kuntzemüller, P. Dhez, F. Wuilleumier and E. Källne, Phys. Rev. A13, 1748 (1976).
20. J. T. Broad and W. P. Reinhardt, J. Phys. B: Atom. Molec. Phys. 8,   (1975).
21. M. Ya Amusya, E. G. Drukarev, V. G. Gorshkov and M. P. Kazachkov, J. Phys. B: Atom. Molec. Phys. 8, 1248 (1975).
22. H. S. Taylor and S. W. Wang, J. Phys. B: Atom. Molec. Phys. 8, 2654 (1976).
23. R. B. Cairns, H. Harrison and R. I. Schoen, Phys. Rev. 183, 52 (1969).
24. Th. M. El-Sherbini and M. J. Van der Wiel, Physica 62, 119 (1972).
25. C. Backx and M. J. Van der Siel, J. Phys. B8, 3020 (1975).
26. G. R. Wight and M. J. Van der Wiel, J. Phys. B, to be published.
27. G. R. Wight, M. J. Van der Wiel and C. E. Brion, J. Phys. B9, 675 (1976).
28. P. Winkler, in preparation.
29. I. B. Bersukev, Opt. Spectrosc. 3, 97 (1957).
    S. Hameed, A. Herzenberg and M. G. James, J. Phys. B1, 882 (1968).
    A. Dalgarno and T. C. Caves, J. Quant. Spectrosc. and Radiat. Transfer 12, 1539 (1972).
30. B. Schneider, Phys. Rev. A7, 557 (1973) and erratum Phys. Rev. A7, 2222 (1973).

Appendix

In the following table we list those systems for which the (γ,2e)-reaction has been investigated either experimentally or theoretically. Contrary to the text we include here also references utilizing the pseudo-photon technique (Ref. 18) and more qualitative discussions of special features (Ref. 16, 17).

Systems	References	
	Experiment	Theory
H^-		20
He	1, 4, 18, 23	2, 3, 21
Li^-		28
Be		28
C		7
Ne	1, 4, 18, 23	5, 6
Ar	1, 4, 18, 23	
Cs^-	11	
Ba	14, 15	16, 17
Kr	23, 24	
Xe	23, 24	
CH_4	25	
NH_3	26	
N_2	27	
CO	27	

THEORETICAL STUDIES OF THE MOMENTUM DISTRIBUTION OF MOLECULAR HYDROGEN*

Vedene H. Smith, Jr.,
Department of Chemistry, Queen's University,
Kingston, Ontario, Canada, K7L 3N6.

ABSTRACT

The theoretical determination of the momentum distribution and Compton profile of $H_2(X^1\Sigma_g^+)$ as a reference for experimental work is discussed with regard to the roles of electron correlation and molecular vibration.

INTRODUCTION

The (e,2e) reaction[1] and Compton scattering[2] of photons and electrons offer promising techniques for the study of electron momentum distributions in atoms and molecules. Inasmuch as both accurate theoretical calculations and experimental measurements are feasible for smaller systems, it is of great importance that such are made in order to probe the validity of the theoretical approximations used to relate the experimental data to the electron momentum distributions. It is for this reason that the systems initially investigated by all three techniques were atomic helium[3-5] and molecular hydrogen[4,6,7]. Whereas good agreement was found for helium, the measurements of the momentum distribution of H_2 by all three techniques were in definite disagreement with the theoretical calculations. The contradictory results for these two-electron systems led to speculation[8] about a possible failure of the Born-Oppenheimer approximation for the molecule. However, a new theoretical calculation and measurement with the (e,2e) reaction were in good agreement[8]. As a result, one must consider the possibility[8] that the reported discrepancies for the Compton profiles could be due to the failure of the impulse approximation (photon scattering [3,9]) and the binary encounter condition (high energy electron scattering, HEES[4]). One would hope that such a failure would possibly be ruled out because of the good agreement obtained for helium.

The situation in early 1976 is summarized in Table I for the peak value $J(0) = \frac{1}{2}\langle p^{-1}\rangle$ of the Compton profile. It is seen that the two experimental measurements listed are essentially the same while the theoretical calculations[10] lie outside the experimental error bounds. Since the calculation which has the energetically best wavefunction gives the closest agreement with the experiments, one

---

*Research supported in part by the National Research Council of Canada.

Table I - Values of $J(0) = \frac{1}{2}\langle p^{-1}\rangle$

Wavefunction	$J(0)$	$-E$
H + H	1.698^{10}	1.00000
LCAO-MO (Optimized Minimal Basis)	1.541^{11}	1.12819
SCF Limit[13]	1.553^{10}	1.13363
Liu CI[12]	1.532^{10}	1.17370
Experiment	$J(0)$	
X-ray[6]	$1.513 \pm 0.7\%$	
HEES[4]	$1.516 \pm 1\%$	

must consider the possibility that a further improvement in the quality of the wavefunction would remove the discrepancy. Such a hope is based upon the (non-rigorous) argument that as the energy improves and the kinetic energy ($\frac{1}{2}\langle p^2\rangle$) increases, $J(0)=\frac{1}{2}\langle p^{-1}\rangle$ should decrease[10]. However, a new measurement (HEES) by Lee[14] yielded $J(0)=1.545 \pm 0.7\%$. This lies between the SCF and correlated (CI) calculations both of which lie just outside the error bounds of this new experiment. As seen from the table, Lee's measurement and the theoretical value calculated[11] for the minimal basis L.C.A.O.-M.O. wavefunction with optimized exponent are in near agreement. It is clear that further theoretical calculations are needed in order to assess the role of electron correlation.

All of these calculations were made with electronic wavefunctions with the internuclear separation being that of the experimental equilibrium value of 1.4 bohrs. Ulsh, Bonham and Bartell[15] suggested that vibrational effects could be important. By means of an expansion technique, they estimated the vibrational correction to a theoretical profile to raise the peak value by approximately 1%. Using better quality wavefunctions and a similar expansion technique, Whangbo, Smith and Clementi[16] came to a similar conclusion. Such a correction would either increase the discrepancy between theory and the older experiments or remove it when the new measurement of Lee is compared with the value for the Liu CI function. However, Braun-Keller and Epstein[17] in a very recent study of vibrational effects using harmonic oscillator and Morse vibrational wavefunctions concluded that the vibrational effect is negligible in contradiction to the results of Ulsh et al[15].

In the present report, we wish to discuss new results on the effects of electron correlation and molecular vibration.

ELECTRON CORRELATION

The Liu wavefunctions used in our earlier study[10] were SCF and correlated ones constructed from linear combinations of atomic orbitals (L.C.A.O.) where the atomic functions were Slater-type orbitals centered on the two nuclei. The correlated wavefunctions involved 39 natural configurations. Liu and Smith[18] have recently made new calculations using both SCF and CI wavefunctions built up from elliptical basis functions (EBF) in order to shed some light upon the role that the type of basis plays in predicting momentum space properties. The correlated EBF wavefunctions were the natural orbital (NO) CI functions of Hagstrom and Shull[19] and Davidson and Jones' NO expansion[20] of the Kolos-Roothaan[21] wavefunction, truncated to 10 natural orbitals. The J(0) values for these two wavefunctions, 1.531 and 1.533, are essentially identical with that listed for the Liu CI wavefunction in Table I., 1.529. In addition, J(0) for a slightly different Liu CI function[22] is 1.532[11,23]. The excellent agreement of all these calculations leads to some confidence in the accuracy of the theoretical value.

To complete the study using EBF wavefunctions, Liu and Smith[18] considered an SCF function constructed with the same EBF basis as the CI function[20]. The J(0) value obtained, 1.558, is in essential agreement with that[10] for the Liu SCF-LCAO function, 1.553.

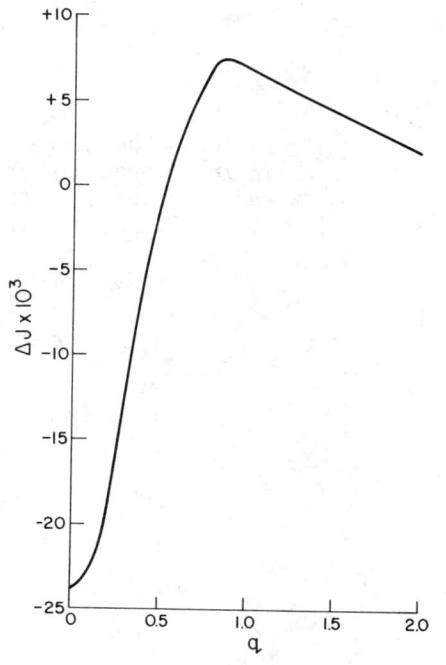

Fig. 1 - The correlation correction $\Delta J(q) = J(q)_{CI} - J(q)_{SCF}$.

Although the above discussion has focused on $J(0)$, we show in Fig. 1, for $0 \leq |q| \leq 2$, $\Delta J(q) = J(q)_{CI} - J(q)_{SCF}$, for these wavefunctions. The effect of electron correlation is seen to be most pronounced in the vicinity of $q = 0$ where the profile is lowered.

MOLECULAR VIBRATION

In view of the disparate views of the role of molecular vibration mentioned in the introduction, Smith et al[23] computed the profile for the Liu CI wavefunctions[22] at 22 different values of the internuclear separation. The profiles $J(q,R)$ were then rigorously averaged over vibrational and rotational wavefunctions computed by direct numerical solution of the radial Schrödinger equation utilizing a nearly exact potential curve[24] which included adiabatic and relativistic corrections to the calculations of Kolos and Wolniewicz[25]. The resulting profile peak for the ground vibrational and rotational state, 1.547, is in excellent agreement with Lee's new measurement, 1.545. We show in Fig. 2 the difference

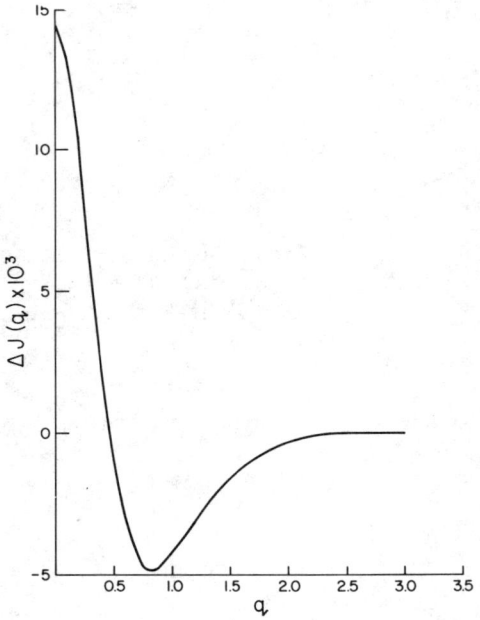

Fig. 2 - The correction in the ground vibrational and rotational state $(V=0, J=0)$ $\Delta J(q)_{00} = J(q)_{00} - J(q, R_e)$

between $J(q)_{V=0,J=0}$ and $J(q,R_e)$ for $0 \leq q \leq 3.5$. Just as in the case of electron correlation, the effect of molecular vibration is seen to be most pronounced at $q=0$ where the peak is raised. Thus, the two effects cancel to a large extent.

Smith et al[23] considered D_2 as well. Comparison of their values of $J(\overline{0})$ in the ground vibrational-rotational state for D_2 (1.543) and H_2 (1.547) indicates a 0.27% lowering due to the isotope effect which is smaller than the effect due to vibrational and rotational averaging.

CONCLUDING REMARKS

The results presented here confirm the accuracy of our earlier calculations as well as the importance of electron correlation and molecular vibration. It appears that Lee's re-measurement using HEES has removed the discrepancy between theory and experiment. New measurements using X-ray or γ-ray scattering are clearly needed.

ACKNOWLEDGEMENTS

The research reported here involved the collaboration of R.E. Brown, E. Clementi, W.H. Henneker, B. Liu, J.W. Liu, A.J. Thakkar and M.H. Whangbo which is gratefully acknowledged. The author would also like to thank the organizers of this conference for their kind invitation which made his participation possible and especially Professor R.A. Bonham for his constant encouragement to resolve this problem.

REFERENCES

1. I.E. McCarthy and E. Weigold, Phys. Reports 27C, 275 (1976).
2. B.G. Williams (editor), Compton Scattering: The Investigation of Electron Momentum Distributions (McGraw-Hill, London, 1977).
3. P. Eisenberger and P.M. Platzman, Phys. Rev. A2, 415 (1970).
4. H.H. Wellenstein and R.A. Bonham, Phys. Rev. A7, 1568 (1973). R.C. Ulsh, H.F. Wellenstein and R.A. Bonham, J. Chem. Phys. 60, 103 (1974).
5. I.E. McCarthy, A. Ugbabe, E. Weigold and P.J.O. Teubner, Phys. Rev. Lett. 33, 459 (1974).
6. P. Eisenberger, Phys. Rev. A2, 1678 (1970).
7. E. Weigold, S.T. Hood, I.E. McCarthy and P.J.O. Teubner, Phys. Lett. 44A, 531 (1973).
8. S. Dey, I.E. McCarthy, P.J.O. Teubner and E. Weigold, Phys. Rev. Lett. 34, 782 (1975).
9. L.B. Mendelsohn and V.H. Smith, Jr., in ref. 2.
10. R.E. Brown and V.H. Smith, Jr., Phys. Rev. A5, 140 (1972).
11. M.H. Whangbo and V.H. Smith, Jr., to be published.
12. B. Liu, unpublished, cited in ref. 10.
13. P.E. Cade and A.C. Wahl, At. Data Nucl. Data Tables 13, 339 (1974).

14. J.S. Lee, Ph.D. Thesis, Indiana University, 1976; J. Chem. Phys. $\underline{00}$, 0000 (1977).
15. R.C. Ulsh, R.A. Bonham and L.S. Bartell, Chem. Phys. Lett., $\underline{13}$, 6 (1972).
16. M.H. Whangbo, V.H. Smith, Jr., and E. Clementi, to be published.
17. E. Braun-Keller and I.R. Epstein, Chem. Phys. Lett. $\underline{40}$, 215 (1976).
18. J.W. Liu and V.H. Smith, Jr., to be published.
19. S. Hagstrom and H. Shull, Rev. Mod. Phys. $\underline{35}$, 624 (1963).
20. E.R. Davidson and L.L. Jones, J. Chem. Phys. $\underline{37}$, 2966 (1962).
21. W. Kolos and C.C.J. Roothaan, Rev. Mod. Phys. $\underline{32}$, 219 (1960).
22. B. Liu, J. Chem. Phys. $\underline{58}$, 1925 (1973).
23. V.H. Smith, Jr., A.J. Thakkar, W.H. Henneker, J.W. Liu, B. Liu, and R.E. Brown, to be published.
24. D.M. Bishop and S. Shih, J. Chem. Phys. $\underline{64}$, 162 (1976).
25. W. Kolos and L. Wolniewicz, J. Chem. Phys. $\underline{41}$, 3663 (1964); $\underline{43}$, 2429 (1965); $\underline{49}$, 404 (1968); Chem. Phys. Lett. $\underline{24}$, 457 (1974); J. Mol. Spectrosc. $\underline{54}$, 303 (1975).

INVESTIGATION OF (e,2e) KNOCKOUT REACTIONS VIA MOLECULAR ELECTRONIC STRUCTURE CALCULATIONS

Geoffrey R.J. Williams
Department of Theoretical Chemistry
The University of Sydney
Sydney, N.S.W. 2000.
AUSTRALIA

ABSTRACT

This paper outlines techniques that may be employed in molecular electronic structure calculations for the investigation of (e,2e) knockout reactions of molecules. One of these methods (the Green function technique) is used to illustrate how the momentum-space wavefunction for the water molecule may be computed.

1. INTRODUCTION

The theoretical investigation of simple knockout reactions for molecular systems may take a number of possible approaches. One such approach is to carry out separate electronic structure calculations for both the target and final products, and from an analysis of the overlap function between the target and products and the computed energy differences, the parameters required for understanding the knockout reaction may be obtained. The particular knockout reaction we will be considering in this article is the (e,2e) reaction for molecular systems:[1]

$$M_o(e,2e)M_s^+$$

where M_o is the target molecule in its ground electronic state o, and M_s^+ is the resultant positive ion of M in the electronic state s.

In order to determine the necessary information regarding the momentum space wavefunction of M_o, we require the overlap function (F/G) between M_o and M_s^+, and the separation energy $\varepsilon = E(M_s^+) - E(M_o)$. As indicated above, these results may be obtained by first carrying out a calculation on the target M_o (which we will assume to have a closed shell electronic structure in this article) followed by a separate open shell calculation for the ion M_s^+. The overlap function may then be computed using standard techniques, and from the difference in total energy between M_o and M_s^+ the separation (or ionisation) energy may be obtained. Details of this type of approach are described in section (2) of this article.

The technique outlined above suffers from a number of disadvantages. Firstly, at least two separate molecular electronic structure calculations are required (or more, if a number of final ion states are to be investigated), and secondly, it is difficult to determine the change in correlation energy that will exist between a closed shell target system and an open ion system, if the two calculations are done in an independent manner such as this. Therefore, it is often preferred to employ more direct methods for determining overlap functions and energy differences between N and N-1 electron systems. Methods that do allow the direct computation of energy differences and overlap functions are the so-called many-body techniques.[2] The two many-body approaches that will be discussed in this article are the equations of motion method,[3] and the closely related Green's function technique.[4] These are discussed in section (3) of this article.

2. DIRECT CALCULATION OF TARGET AND ION WAVEFUNCTIONS

The wavefunctions for the target system with N electrons (which we are assuming to be the ground state of a closed shell molecule) may be written as the Slater determinant:[5]

$$\Psi = |\phi_1 \bar{\phi}_1 \phi_2 \bar{\phi}_2 \cdots \phi_N \bar{\phi}_N| \qquad (1)$$

where $\phi_i = \alpha \phi_i$ and $\bar{\phi}_i = \beta \phi_i$

The variational theorem is then used to find the best set of one-electron functions for the use in determinant (1), by minimising the energy E, while constraining the functions to remain orthonormal. This leads to the set of interigo-differential equations:

$$F_i \phi_i = E_i \phi_i$$

The effective hamiltonian F_i for an electron in ϕ_i depends on the position of all the other electrons through their average coulomb and exchange fields. This, of course, is the well-known Hartree-Fock method, and the operator F is often referred to as the Fock operator.

Hall and Roothaan developed a method for obtaining solutions to the Hartree-Fock equations which avoids numerical integrations, by expanding the orbitals $\{\phi\}$ in terms of a complete set of basis functions $\{\chi\}$.

$$\phi_i = \sum_p \chi_p C_{pi}$$

or in matrix notation $\phi = \chi C$
where C is the matrix of coefficients.
The basis functions $\{\chi\}$ most commonly employed in molecular calculations are either
  (a) Slater-type orbitals
$$\chi(d) = [(2N!)]^{-\frac{1}{2}}(2d)^{N+\frac{1}{2}} r^{N-1} e^{-\alpha r} Y_{\ell m}(\theta, \phi)$$

or, (b) Gaussian functions

$$\chi(\alpha) = [(2N-1)!!]^{-\frac{1}{2}}(2\pi)^{-\frac{1}{4}}r^{N-1}e^{-ar^2}Y_{\ell m}(\theta,\phi)$$

Although the Gaussion functions are less "atomic like" than Slater functions (which generally means that one requires many more Gaussions than Slaters to form a reasonable basis), the fact that multi-centre two-electron are much easier to solve over Gaussion functions, makes them the most popular functions for molecular calculations.[6]

In terms of a given basis, the Hall-Roothaan equations become:

$$\underline{F}\,\underline{C} = \underline{S}\,\underline{C}\,\underline{E}$$

where:
$\underline{C}$ coefficient matrix
$\underline{E}$ orbital energy matrix
$\underline{S}$ overlap matrix
$\underline{F}$ Hartree-Fock hamiltonian matrix

$$\underline{F} = \underline{H} + \underline{G}$$

$$H_{ut} = <u|-\tfrac{1}{2}\nabla^2|t> + <u|-\sum_A \frac{Z_A}{r_A}|t>$$

$$G_{ut} = \sum_{r,s} R_{rs}[2(ut|rs) - (ur|ts)]$$

$$(ut|rs) = \int\int \chi_u(1)\chi_t(1) r_{12}^{-1}\chi_r(2)\chi_s(2)\,d\tau_1 d\tau_2$$

$$<u|\Theta|t> = \int \chi_u(1)\Theta(1)\chi_t(1)d\tau_1$$

Matrix $2\underline{R} = \underline{P}$ is the matrix representation of the first order density matrix in the χ-basis, and has elements defined by

$$R_{rs} = \sum_{occ} C_{ri} C_{si}$$

The total electronic energy is given by $E = tr(\underline{R}\,\underline{H}) + tr(\underline{R}\,\underline{F})$
And the total energy is $E = tr(\underline{R}\,\underline{H}) + tr(\underline{R}\,\underline{F}) + \sum_{A<B}\frac{Z_A Z_B}{R_{AB}}$

The wavefunction for the ion will be generally more difficult to compute, than for the target, because it is an open shell system. There are a large number open shell techniques that may be employed to compute the wavefunction for the ion, but we shall only consider one representative technique in this article. The method we will discuss in this article is the unrestricted Hartree-Fock (UHF) method.[7]

A UHF wavefunction has the form:

$$\Psi = \det\{\upsilon_1 \upsilon_2 \ldots \upsilon_p \upsilon_{p+1} \upsilon_{p+2} \ldots \upsilon_N\}$$

where p is the number of α-electrons
and   q is the number of β-electrons
and   N = p + q

The functions υ and ν form two <u>different</u> orthonormal sets

$$\upsilon_i = \sum_s \chi_s a_{si} \qquad \nu_i = \sum_s \chi_s b_{si}$$

The total electronic energy is given by:

$$E = tr(\underline{P^\alpha H}) + tr(\underline{P^\beta H}) + \tfrac{1}{2} tr(\underline{P^\alpha G^\alpha}) + \tfrac{1}{2} tr(\underline{P^\nu G^\beta})$$

where:

$$H_{st} = <s|-\tfrac{1}{2}\nabla^2 - \Sigma \frac{Z_A}{r_A}|t>$$

$$G^\alpha_{st} = \sum_{uv} [P_{uv}(st|uv) - P^\alpha_{uv}(su|tv)]$$

$$G^\beta_{st} = \sum_{uv} [P_{uv}(st|uv) - P^\beta_{uv}(su|tv)]$$

and $\quad P^\alpha_{st} = \sum_i^p a_{si} a_{ti} \; : \; P^\beta_{st} = \Sigma b_{si} b_{ti}$

Minimising the energy, subject to the usual orthonormality constraints, results in the following equations

$$\underline{F^\alpha}\,\underline{a} = \underline{S}\,\underline{a}\,\underline{E}$$

$$\underline{F^\beta}\,\underline{b} = \underline{S}\,\underline{b}\,\underline{E}$$

where $\quad \underline{F^\alpha} = \underline{H} + \underline{G^\alpha}$

$\qquad\qquad \underline{F^\beta} = \underline{H} + \underline{G^\beta}$

These equations are solved by iteration until self-consistency is obtained.

The UHF wavefunction is not generally an eigenfunction of S^2, but contains components of several spin states:

$$\Psi_{UHF} = \sum_{m=0}^{q} C_{s'+m} \Psi_{s'+m}$$

$S' = \tfrac{1}{2}(p-q) \quad \Psi_{s'+m}$ is an eigenfunction of S^2 with multiplicity $2(s'+1)+1$.

To obtain a wavefunction of pure spin symmetry S' one may apply a projector operator $\hat{O}_{s'}$ to the UHF wavefunction.

$$\hat{O}_{s'} \Psi_{UHF} = \Psi_{s'}$$

where $\hat{O}_{s'} = \prod_{k \neq s'} \{ \hat{A}_k / [S'(S'+1) - k(k+1)] \}$

and $\hat{A}_k = \hat{S}^2 - k(k+1)$, and is the annihilation operator which removes the component of multiplicity $k(k+1)$ from the wave function Ψ_{UHF}.

SINGLE ANNIHILATION

It is often found that single annihilation provides a sufficient approximation to full projection: [8]

$$\hat{A}_{s'+1} \Psi_{UHF} = \Psi_{AUHF} \simeq \Psi_{s'}$$

The energy after single annihilation is obtained by evaluating

$$E = \langle H \rangle = \frac{\int (\hat{A}_{s'+1} \Psi_{UHF})^* \hat{H} (\hat{A}_{s'+1} \Psi_{UHF}) dx}{\int (\hat{A}_{s'+1} \Psi_{UHF})^* (\hat{A}_{s'+1} \Psi_{UHF}) dx}$$

Hence the projected unrestricted wavefunction and energy may be obtained by this procedure. Full details may be found in reference.[8]

Having obtained the total energies for the target and ion, it is now necessary to obtain the overlap function (F/G) between the two, so that the momentum information may be obtained. Because the wavefunctions for the target and ion have been computed independently, they will be, in general, non-orthogonal to one another. An efficient procedure for obtaining the overlap between two non-orthogonal determinental functions is by use of the corresponding orbital transformation.[9]

CORRESPONDING ORBITAL TRANSFORMATION

Consider the wavefunction for the ion $\underline{\Psi}_A$ and target Ψ_B.

$$\Psi_A = \det \{a_1(1) a_2(2) \ldots \ldots a_N(N)\}$$
$$\Psi_B = \det \{b_1(1) b_2(2) \ldots \ldots b_M(M)\}$$

where $M = N + 1$

let $\underline{a} = (a_1 a_2 \ldots a_N)$

$\underline{b} = (b_1 b_2 \ldots b_M)$

There exists a non-orthogonality between the sets:

$$\underline{D} = \int \underline{b}^+ \underline{a} \, d\tau$$

Consider the unitary transformations

$$\underline{\hat{a}} = \underline{a} \, \underline{v}$$

$$\underline{\hat{b}} = \underline{b} \, \underline{u}$$

which leave Ψ_A and Ψ_B invarient except in phase

$$\Psi_A = \det(\underline{v}^+) \det\{\hat{a}_1(1)\hat{a}_2(2)\ldots\hat{a}_N(N)\}$$
$$\Psi = \det(\underline{u}^+) \det\{\hat{b}_1(1)\hat{b}_2(2)\ldots\hat{b}_M(M)\}$$

and transform the overlap matrix:

$$\int \underline{\hat{b}}^+ \underline{\hat{a}} \, d\tau = \underline{\hat{d}}(\text{diagonal})$$

$$\underline{\hat{d}} = \underline{u}^+ \underline{D} \, \underline{v}$$

multiply the equation above by its adjoint

$$\underline{v}^+ \underline{D}^+ \underline{u} = \underline{\hat{d}}^*$$

$$\underline{u}^+ \underline{D} \, \underline{v} \, \underline{v}^+ \underline{D}^+ \underline{u} = \underline{\hat{d}} \, \underline{\hat{d}}^*$$

but $\underline{v} \, \underline{v}^+ = 1$

$$\underline{u}^+ (\underline{D} \, \underline{D}^+) \underline{u} = \underline{\hat{d}} \, \underline{\hat{d}}^*$$

i.e. $\underline{u}$ may be found by diagonalisation of Hermitian matrix $\underline{D} \, \underline{D}^+$

Similarity may show that $\underline{v}$ may be found by

$$\underline{v}^+ (\underline{D}^+ \underline{D}) \underline{v} = \underline{\hat{d}}^* \, \underline{\hat{d}}$$

hence the overlap matrix becomes:

$$S_{AB} = \int \Psi_B^* \, \Psi_A \, d\tau$$

$$(F/G) = S_{AB} = \{\det(\underline{u}) \det(\underline{v}^+) \prod_{i=1}^{N} d_{ii}\} \, \hat{b}_M$$

which is readily computed.

3. APPLICATION OF MANY-BODY TECHNIQUES TO THE CALCULATION OF SEPARATION ENERGIES AND OVERLAP FUNCTIONS

We now wish to examine techniques which allow a more direct calculation of separation energies and overlap functions, and which avoid some of the problems associated with carrying out separate calculations on target and ion states.

The first method we will discuss is the Equations of Motion method.

THE EQUATIONS OF MOTION (EOM)

Following Rowe's[3] variational formulation of the Heisenberg equation of motion, McKoy[*] and others have developed approximate solutions that proved very successful in the calculation of electronic excitation energies and oscillator strengths in molecules. Simons and Smith further adapted the EOM method to deal with electron attachment and ionisation processes and since then Simons and co-workers [14] have demonstrated on a number of molecules the accuracy of their approach. As the above reference contains all the basic theory, we shall restrict ourselves here and present only a brief description of the method.

Given the exact neutral molecule ground state wavefunction $|g\rangle$, the exact eigenstate $|\mu^+\rangle$ of the positive ion is obtained through the excitation operator Ω_μ as follows:

$$|\mu^+\rangle = \Omega_\mu |g\rangle$$

Assuming that $|g\rangle$ and $|\mu^+\rangle$ obey the Schrödinger equations

$$H|g\rangle = E|g\rangle$$

and
$$H|\mu^+\rangle = E_\mu^+ |\mu^+\rangle,$$

the following basis equations are obtained,

$$[H,\Omega_\mu]|g\rangle = (E_\mu^+ - E)\Omega_\mu |g\rangle$$
$$\Delta E_\mu \, \Omega_\mu |g\rangle \qquad (2)$$

where ΔE_μ is the μ-th ionisation potential.
The operator Ω_μ is now approximated in terms of the set of second quantised Hartree-Fock orbital creation $\{C_i^+\}$ and annihilation $\{C_i\}$ operators as follows:

$$\Omega_\mu = \sum_i X_i(\mu) C_i^+ + \sum_{m<n,\alpha} Y_{n\alpha m}(\mu) C_n C_\alpha^+ C_m$$
$$+ \sum_{\alpha<\beta,m} Y_{\alpha m}(\mu) C_\alpha C_m^+ C_\beta,$$

where α,β label occupied Hartree-Fock spin orbitals, m,n label both sets.

$\{X_i(\mu)\}$, $\{Y_{n\alpha m}(\mu)\}$ and $\{Y_{\alpha m\beta}(\mu)\}$ are the sets of variational coefficients to be determined.

$|g\rangle$ is now approximated as a configuration interaction wavefunction containing the Hartree-Fock single determinant plus all double excitations the co-efficients of which are obtained by Rayleigh-Schrodinger perturbation theory, i.e.

$$|g\rangle = N[|0\rangle + \sum_{m<n,\alpha<\beta} \left(\frac{\langle mn||\alpha\beta\rangle}{\varepsilon_\alpha + \varepsilon_\beta - \varepsilon_m - \varepsilon_n} \right) C_m^+ C_N + C_\beta C_\alpha |0\rangle]$$

where N is a normalisation constant,

$|0\rangle$ is the Hartree-Fock single determinant,

$\{\varepsilon_i\}$ are the set of orbital energies and

$\{\langle mn||\alpha\beta\rangle\}$ are the set of "antisymmetrised" two-electron integrals over spin orbitals

(i.e. $|mn||\alpha\beta\rangle = \langle m(1)n(2) r_{12}^{-1}(1-P_{12}) \alpha(1)\beta(2)\rangle$)

Substituting the above expressions for Ω_μ and $|g\rangle$ into equation (2) and casting it into variational form yields the matrix equations.

$$\underline{\underline{A}} \, \underline{X}(\mu) + \underline{\underline{B}} \, \underline{Y}(\mu) = \Delta E_\mu \, \underline{X}(\mu)$$

$$\underline{\underline{B}}^+ \, \underline{X}(\mu) + \underline{\underline{D}} \, \underline{Y}(\mu) = \Delta E_\mu \, \underline{\underline{S}}(\mu)\underline{Y}(\mu)$$

Formally solving equation for $\underline{Y}$ and substituting it in, yields our final working equation which has the pseudo-eigenvalue form:

$$\underline{\underline{H}}(\Delta E_\mu) \underline{X}(\mu) = \Delta E_\mu \underline{X}(\mu) \tag{3}$$

The full expressions for $\underline{\underline{H}}(\Delta E_\mu)$, $\underline{\underline{A}}$ and $\underline{\underline{B}}$ are listed in ref. 14. With their use, equation (2) is obtained; this is valid to third order with respect to the electronic interaction r_{ij}^{-1}, and is solved iteratively.

Solution of this equation provides the ionisation energies ΔE and the coefficient vectors $\underline{X}$ and $\underline{Y}$ from which the overlap function may be evaluated.

A closely related technique to the equations of motion is the Green's function method. The poles of the one-particle propagator (Green's function) are directly associated with the ionisation potentials and electron affinities of the target systems.

However, in this article we will concentrate on the calculation of the overlap function, from the one particle propagator (rather than the ionisation potentials) and illustrate its use in the case of the water molecule.[13]

CALCULATION OF THE GENERALISED OVERLAP AMPLITUDE VIA MANY-BODY TECHNIQUES

The Fourier transform of the one-particle propagator obeys the equation of motion [4,13]

$$G_{pq}(E) = \langle\langle a_p^\dagger ; a_q \rangle\rangle_E$$

$$= E^{-1}\{\langle [a_p^\dagger, a_q]_+ \rangle + \langle\langle a_p^\dagger ; [a_q, H]_- \rangle\rangle_E\}$$

where the annihilation and creation operators (a_q and $a_p^\dagger$) obey the anticommutation relations:

$$[a_p^\dagger, a_q]_+ = a_p^\dagger a_q + a_q a_p^\dagger = \delta_{pq}$$

and H is the hamiltonian written in second quantised form:

$$H = \sum_{pq} \langle p|h_1|q\rangle a_p^\dagger a_q + \tfrac{1}{2} \sum_{pqrs} \langle pq|h_{12}|sr\rangle a_p^\dagger a_q^\dagger a_r a_s.$$

The spin orbitals p, q etc. form an orthonormal set. We consider the coordinate space representation of these orbitals.

It is convenient to introduce the field operator

$$\phi(1) = \sum_p a_p p(1)$$

associated with the complete set of spin-orbitals $\{p(1)\}$. It is thus possible to define a Green's function:

$$G(1,1';E) \equiv \langle\langle \phi^\dagger(1) ; \phi(1') \rangle\rangle_E$$

$$= \sum_{pq} p(1)^* q(1') \langle\langle a_p^\dagger ; a_q \rangle\rangle_E$$

which has the spectral resolution:

$$G(1,1';E) = \sum_s \left[\frac{g_s(1) g_s^*(1')}{E + E_s^{(N-1)} - E_0^{(N)}} + \frac{f_s(1) f_s^*(1')}{E + E_0^{(N)} - E_s^{(N+1)}} \right]$$

It is evident that this expression has poles at $E = E_0^{(N)} - E_s^{(N-1)}$

associated with ionisation potentials, and at $E = E_s^{(N+1)} - E_o^{(N)}$ associated with electron affinities.

The generalised overlap amplitudes for the target ground state 0 and the final ion state s

$$g_s(1) = \langle \psi_s^{(N-1)} | \phi(1) | \psi_o^{(N)} \rangle \equiv (F | G),$$

$$f_s(1) = \langle \psi_o^{(N)} | \phi(1) | \psi_s^{(N+1)} \rangle$$

do not form an orthonormal set, but have physical significance as they are directly associated with the elementary process of ionisation or electron attachment. They are also related to the natural orbitals of the N-electron system via a canonical transformation.[10]

Following Pickup and Goscinski[4] by considering the set of spin-orbital electron field operators $\underline{a} = \{a_s\}$, we can write

$$\underline{G}(E) = (\underline{a} | (E\hat{1} - \hat{H})^{-1} | \underline{a}). \qquad (4)$$

Approximating the superoperator resolvent $(E\hat{1} - \hat{H})$ via inner projection techniques[4] yields:

$$\underline{G}(E) = (\underline{a} | \underline{h})(\underline{h} | E\hat{1} - \hat{H} | \underline{h})^{-1}(\underline{h} | \underline{a}),$$

where $\underline{h}$ is a projection manifold from our operator space of Fermion-like operators. A particular choice of $\underline{h}$ consists of the union of the two subspaces[4]

$$\underline{h}_1 = \{a_i\}, \{a_a\} \text{ for } i < N < a,$$

$$\underline{h}_3 = \{a_a^\dagger a_i a_j\}, \{a_i^\dagger a_a a_b\} \text{ for } i < j < N < a < b,$$

where the spin-orbital field operators are associated with a Hartree-Fock basis and the following conventions have been adopted: i, j denote hole states, c denotes the characteristic orbital, i.e. the one from which an electron is ejected, a, b denote particle states and p, q denote unspecified spin-orbitals. These specifications of the subscripts guarantee that the basis will be linearly independent. When the N-electron ground state average is restricted to the HF ground state, we obtain:

$$(\underline{a} | \underline{h}_3) = (\underline{h}_1 | \underline{h}_3) = \underline{0},$$

$$(\underline{h}_1 | \underline{h}_1) = \underline{1} \text{ and } (\underline{h}_3 | \underline{h}_3) = \underline{1},$$

and the following expression is readily derived from partitioning of matrices:

$$\underline{G}^{-1}(E) = (\underline{h}_1 | E1 - H | \underline{h}_1) - (\underline{h}_1 | H | \underline{h}_3)(\underline{h}_3 | E1 - H | \underline{h}_3)^{-1}(\underline{h}_3 | H | \underline{h}_1) \qquad (5)$$

In the HF basis

$$(h_1|E\hat{1}-\hat{H}|h_1)^{-1} = \frac{\delta_{pq}}{E-\epsilon_p}$$

The simplest non-trival approximation to equation (5) is obtained by replacing $\hat{H}$ in the inverse matrix $(h_3|E\hat{1}-\hat{H}|h_3)^{-1}$ by the Fock superoperator $\hat{F}$, which is defined by the equation

$$\hat{F} X = [X, F]$$

$$F = \Sigma \epsilon_p a_p^\dagger a_p ,$$

where the ϵ_p are HF eigenvalues.

For the general spin-orbitals p and q we then obtain

$$G^{-1}(E)_{pq} = (E-\epsilon_p)\delta_{pq} - \frac{1}{2} \sum_{iab} \frac{\langle ab||pi\rangle\langle qi||ab\rangle}{E+\epsilon_i-\epsilon_a-\epsilon_b}$$

$$-\frac{1}{2} \sum_{ija} \frac{\langle ij||pa\rangle\langle qa||ij\rangle}{E+\epsilon_a-\epsilon_i-\epsilon_j}$$

where the two-electron molecular integrals are defined as

$$\langle pq||rs\rangle = \int\int p^*(1)q^*(2)h_{12}(1-P_{12})r(1)s(2)d^3r_1 d^3r_2 ,$$

and P_{12} permutes the coordinates of electrons 1 and 2.

At this stage we make the approximation of associating a generalised overlap amplitude with a particular characteristic orbital c. The overlap amplitude g_{cs} associated with the ionisation of orbital c to the ion state s may be expanded in terms of the complete set of spin orbitals

$$g_{cs}(1) = \langle \psi_{cs}^{(N-1)}|\phi(1)|\psi_o^{(N)}\rangle$$

$$= \sum_p C^s_{cp} p(1).$$

The coefficients appearing in the expression above may be obtained perturbationally since:

$$g^*_{cs}(1)g_{cs}(1') = \frac{1}{2\pi i} \int G(1,1',\omega)d\omega$$

$$= \frac{1}{2\pi i} \int_{\rho_s} \Sigma\, p^*(1)q(1')G(\omega)d\omega,$$

where ρ_s is a contour enclosing only the pole ε_s.

In the case where the many correlated states ψ_{cs} with the same symmetry as the orbital c are dominated in a localised energy region by one state who coefficient C^s_{cc} is large, we perform the contour integration round the nearby Koopmans pole ε_c. If we expand the expression for the Green's function (equation 4) about the Koopmans poles, we obtain the expression:

$$G_{pq}(\omega) = \frac{\delta_{pq}}{\omega\,\varepsilon_p} + \frac{1}{\omega\,\varepsilon_p}\left[\frac{1}{2}\sum_{ija}\frac{\langle ij||pa\rangle\langle aq||ij\rangle}{\omega-\varepsilon_i\,\varepsilon_j\,\varepsilon_a}\right.$$

$$\left.+ \frac{1}{2}\sum_{abi}\frac{\langle ab||pi\rangle\langle qi||ab\rangle}{\omega-\varepsilon_a-\varepsilon_b+\varepsilon_i}\right]\frac{1}{\omega\,\varepsilon_q}$$

$$+ \text{ higher order terms,}$$

which may be integrated term by term along ρ. The residues occurring at the poles $\omega = \varepsilon_p\,\varepsilon_q$ are

$$\text{Res}(\varepsilon_p) = 1 - \frac{1}{2}\sum_{ija}\frac{|\langle ij||pa\rangle|^2}{(\varepsilon_p+\varepsilon_a-\varepsilon_i-\varepsilon_j)^2}$$

$$-\frac{1}{2}\sum_{abi}\frac{|\langle ab||pi\rangle|^2}{(\varepsilon_p+\varepsilon_i-\varepsilon_a-\varepsilon_b)^2} \quad \text{for } p = q$$

$$= \frac{1}{2}\sum_{ija}\frac{\langle ij||pa\rangle\langle qa||ij\rangle}{(\varepsilon_p+\varepsilon_a-\varepsilon_i-\varepsilon_j)(\varepsilon_p-\varepsilon_q)}$$

$$+ \frac{1}{2}\sum_{abi}\frac{\langle ab||pi\rangle\langle qi||ab\rangle}{(\varepsilon_p+\varepsilon_i-\varepsilon_a-\varepsilon_b)(\varepsilon_p\,\varepsilon_q)} \quad \text{for } p \neq q$$

Using the expansion for $G_{pq}(\omega)$ and these residues, we obtain the following expressions for the coefficients C^s_{cp}:

$$C^s_{cc} = 1 - \frac{1}{4}\sum_{ija}\frac{|\langle ij||ca\rangle|^2}{(\varepsilon_c+\varepsilon_a-\varepsilon_i-\varepsilon_j)^2} - \frac{1}{4}\sum_{abi}\frac{|\langle ab||ci\rangle|^2}{(\varepsilon_c+\varepsilon_i-\varepsilon_a-\varepsilon_b)^2}$$

$$C^s_{cp} = \frac{1}{2} \sum_{ija} \frac{\langle ij||ca\rangle\langle pa||ij\rangle}{(\epsilon_c+\epsilon_a-\epsilon_i-\epsilon_j)(\epsilon_i-\epsilon_p)} + \frac{1}{2} \sum_{iab} \frac{\langle ab||ci\rangle\langle pi||ab\rangle}{(\epsilon_c+\epsilon_i-\epsilon_a-\epsilon_p)(\epsilon_c-\epsilon_p)}$$

$$(p \neq c). \qquad (6)$$

If we remove the approximation of associating a generalised overlap amplitude with a particular characteristic orbital c, equation (6) includes a sum over c.

These expressions may also be derived by writing the ground state and ionised state wave functions as configuration-interaction wave functions including all double and single excitations. Hence the expression

$$(F|G\rangle = \sum_p C^s_{cp} p(1)$$

is employed in the many-body computation of the generalised overlap amplitude. This expression takes into account correlation and relaxation effects via the terms appearing in the expressions for the coefficients C^s_{cp} (equation 6).

The equations developed in this section indicate that the generalised overlap amplitudes obtained by contour integration contain rather detailed information, even though the approximations are developed from the Hartree-Fock N-particle ground state. This phenomenon of 'correlation feedback' is an essential feature of the propagator formalism.

In Figure 1 we have shown the experimental values obtained at 400 eV and 1200 eV for the angular correlations of the four valence orbitals for water together with the theoretical profiles computed from:

  (i) The Fourier transforms of the Snyder-Basch [11] molecular orbitals obtained from a (4s,2p/2s) basis (profile A).
  (ii) The Fourier transforms of the Dunning molecular orbitals obtained from the (5s, 3p/3s)[12] basis (profile B), and
  (iii) The Fourier transform of the generalised overlap amplitudes obtained from the (5s, 3p/3s) basis (profile C).

Comparison of profiles A and B shows the effect of increasing the size of the atomic basis set, in particular of adding longer-range orbitals. Much better agreement for the $1b_1$, $3a_1$ and $1b_2$

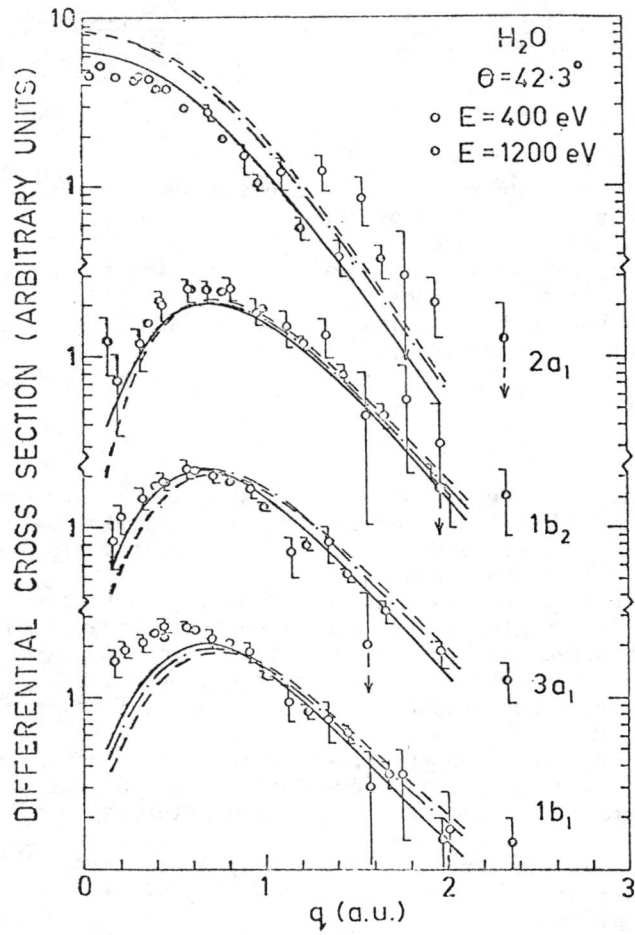

The noncoplanar symmetric (e,2e) differential cross sections at 400 eV(O) for the principal valence orbital transitions in H_2O plotted as a function of the recoil momentum q. The curves show the plane wave calculations using:

(A) the molecular orbital wave functions of Snyder and Basch (----);
(B) Dunning molecular orbitals obtained from the (5s,3p/3s) basis (—.—.) and
(C) the generalised overlap amplitudes obtained from the (5s,3p/3s) basis (———).

The cross sections for the different transitions are determined relative to each other, the absolute magnitude is not measured. The calculated cross sections have the experimental angular (q) resolution folded in. For low q the curves B and C are indistinguishable for the $3a_1$ and $1b_2$ transitions.

orbitals is obtained for low recoil momenta q using the longer-range basis. This means that there is a higher electron density at large distances from the nuclei than allowed by the smaller basis set. For the $1b_1$ orbital, agreement is still not satisfactory at low q. This is a strong indication that still longer-range terms need to be included in the basis set. This is one of the first indications of the value of the (e,2e) reaction in determining details of charge density.

Comparison of profiles B and C gives an indication of the importance of correlation and relaxation effects.

The one-electron generalised overlap amplitude is given by the expression

$$g_{cs}(1) = \sum_p C^s_{cp} p(1)$$

where $\{p(1)\}$ is the set of molecular spin orbitals used as a basis.

In the case of water, this is the complete set of molecular spin-orbitals obtained from the atomic (5s, 3p/3s) basis.

If no relaxation or correlation effects are included, i.e. we employ the target Hartree-Fock approximation and neglect all configuration interaction, then $C^s_{cc} = 1$ and $C^s_{cp} = 0$ (all p) so that

$$g_{cs}(1) \equiv c(1)$$

Hence, in this case, the generalised overlap amplitude is just the spin-orbital C(1) from which the electron was ejected.

In the general case however, $C^s_{cc} \neq 1$ and $C^s_{cp} \neq 0$. In table 1 we have summarised the values obtained for $[C^s_{cc}]^2 = S_{cs}$ for the principal state s of water, that is, the ones with the largest coefficients C^s_{cc} in the neighbourhood of the Koopmans pole ε_c.

It is evident from the table that for the 'outer' valence orbitals for water ($1b_1$, $3a_1$, and $1b_2$) ranging in energy from 14-20 eV, the generalised overlap amplitude consists mainly of the original spin-orbital from which the electron was removed together with a small contribution from other spin-orb itals. However, for the deeper (37ev) $2a_1$ level, the spectroscopic factor is approximately 20% smaller than for the outer levels, indicating an increase in importance of relaxation and correlation effects for this level. This result is consistent with what one might expect for more tightly bound levels, since pair correlation energy increases for more tightly bound electrons and the creation of a hole in a deeper level would be expected to result in greater "shake-up" and relaxation processes, then would the removal of an outer electron.

TABLE 1

Theoretical orbital energies and Spectroscopic factors for the principal ion states of the water molecule.

Orbital	Orbital energy (ε_c)	S_{cs}
$1b_1$	13.90eV	0.868
$3a_1$	15.55eV	0.877
$1b_2$	19.60eV	0.907
$2a_1$	37.20eV	0.714
$1a_1$	559.67eV	0.383

CONCLUSION

In this article we have briefly outlined some of the techniques employed in molecular physics to obtain information regarding the (e,2e) knockout reaction for molecular systems.

REFERENCES

1. I.E. McCarthy and E. Weigold, Physics Reports, $\underline{27C}$, 275 (1976)
2. D.J. Rowe, Rev. Mod. Phys. $\underline{40}$, 153 (1968)
3. T. Shibura, J. Rose and V. McKoy, J. Chem. Phys. $\underline{58}$ 500 (1973)
4. B.T. Pickup and O. Goscinski, Mol. Phys. $\underline{26}$, 1013 (1973)
5. R. McWeeny and B.T. Sutcliffe, Methods of Molecular Quantum Mechanical (Academic Press, 1969, N.Y.)
6. H. Talceta, S. Huzinaga and O-Ohata, J. Phys. Soc. Japan, $\underline{21}$ 2313 (1966)
7. J.A. Pople and R.K. Nesbet, J. Chem. Phys. $\underline{22}$, 571 (1954)
8. T. Amos and L.C. Snyder, T. Chem. Phys. $\underline{41}$, 1773 (1964)
9. H.F. King, R.F. Stanton, H. Kim, R.E. Wyatt and R.G. Parr, J. Chem. Phys. $\underline{47}$, 1936 (1967)
10. O. Goscinski and R. Lindner, J. Math. Phys. $\underline{11}$, 1313 (1970)
11. L.C. Snyder and H. Basch, Molecular Wavefunctions and Properties (Wiley, 1972, N.Y.)
12. T.H. Dunning, J. Chem. Phys. $\underline{55}$, 716 (1971)
13. G.R.J. Williams, I.E. McCarthy and E. Weigold, Chem. Phys. (in the press)
14. T. Simons and W.D. Smith, J. Chem. Phys. $\underline{58}$, 4899 (1973)

NUCLEAR PHYSICS IN AUSTRALIA

B.M. Spicer

School of Physics, University of Melbourne,
Parkville, Victoria, Australia 3052

Some of the research programs in nuclear physics at the Australian Atomic Energy Commission, the Australian National University, and the University of Melbourne are described. These programs involved studies of neutron capture reactions, radiative capture of α-particles, formation of neutron-rich nuclides (the example given is ^{22}O), radiative decay of high spin states in deformed nuclei formed in (heavy ion, xn) reactions, yields of proton capture reactions important to theories of nucleosynthesis, photodisintegration studies of $1f_{7/2}$-shell nuclides, and attempts to observe higher multipole giant resonances in photonuclear reactions in ^{208}Pb.

NUCLEON KNOCKOUT: OFF-SHELL EFFECTS[†]

G. J. Stephenson, Jr.
Los Alamos Scientific Laboratory, Los Alamos, N.M. 87545

I want to discuss some questions which address both the difficulty and the challenge of using nucleon knock-out reactions with hadronic probes to determine momentum space wavefunctions for bound orbitals. There is a difference in principle between these reactions and those which employ photon or electron probes in the study of electron wavefunctions; that difference is the knowledge of the basic interactions between the probe and the target constituents. In fact, I shall endeavor to convince you first that the prospects are indeed impossible in principle given our current knowledge of the strong interactions, and then I shall attempt to convince you that it may well be practically feasible anyway. For this presentation I shall eschew slides and equations since they all have been shown before or can be perused at leisure.

The difference, of course, stems from the fact that we do not have a complete and tested theory like quantum electrodynamics upon which we can base our strong-interaction calculations. Even if one of the candidates being touted is a theory of all interactions, or even just of strong interactions, is correct, the calculation of the nucleon-nucleon interaction from such a theory is far off. We are therefore left to build models which we confront with various forms of data. The most obvious are phase shifts fit to elastic scattering data, including polarization, depolarization and polarization rotation data. One problem which I hope will fast disappear is the fact that, even at low and intermediate energies, the data are not yet complete enough to remove all ambiguities from those fits.

Nevertheless, one builds models with some parameters and fits the data moderately well. As the model becomes more sophisticated, as more information about known mesons is fed in along with feedback from nuclear matter calculations, from nuclear structure calculations and from attempts to fit electromagnetic properties of few nucleon systems, the models eventually earn the sobriquet "realistic." While there are many such models, they generally share the same long range behavior, similar medium range behavior, and differ primarily, albeit often drastically, at distances of less than one-half fermi. If divers potentials give equivalent fits to the data, how then is one to choose?

To illustrate the ambiguity even further, let me describe a little theoretical game which has been in vogue the last several years.[1]) Consider some potential which is deemed to give an adequate representation of the elastic scattering phase-shifts. This means that, in each partial wave, the potential gives the correct asymptotic behavior. Now define a unitary transformation which acts on the relative coordinate of the two nucleons and is short-ranged. By that I mean that the effect of the transformation must vanish faster than 1/r at infinity, although all work of interest deals with transformations which effectively die within a few fermis. Since the wavefunction at infinity is identical in all partial waves, the nucleon-nucleon phase shifts are all identical. However, the potential is clearly changed in its radial form. In this sense, we see that even a perfect knowledge of the phase shifts is not sufficient to determine the short ranged behavior of the nucleon-nucleon system. One does have to do a shuffle with the kinetic energy, but one can in this way define two different potentials which give exactly the same scattering fit.

Where then is there a difference? Redish[2] discussed this somewhat yesterday in terms of off-shell two-body t-matrices in many body scattering theory. Since the half-off-shell partial-wave t-matrix (assuming uncoupled partial waves here) is the Bessel-Fourier transform of the potential times the scattering wavefunction at the on-shell momentum, it will clearly be affected by the transformations described above. In coordinate space this is the statement that the wavefunction of one of an interacting pair of nucleons is affected by a third nucleon near one of the pair. Obviously, whether one views the system in momentum space through a multiple scattering theory or thinks of bound state properties in coordinate space, the many-body properties will be altered by the change of potential.

At this point a comment about the unitary transformations is in order. A truly unitary transformation applied to any system is supposed to leave all of the observables unchanged. However, I just finished claiming that the many-body properties of a system of nucleons does change under what I called unitary transformations above. The point is that I discussed the transformations as if they occurred only in the two-particle subspace of Fock space. In order to properly define a unitary transformation which does what I want in the two-body relative space, I must define its operation on some arbitrary representation of one body states and then exponentiate it. Expanding the exponentials leads to three, four and many-body terms. Thus, if I had started with a "true" two-body interaction fitting the two-body data, the transformation would induce many-body interaction terms, terms which depended on the coordinates of more than two nucleons simultaneously. Evaluating these many-body terms would restore the constancy of the observables no matter how many particles were involved.

As advertised, the problem we posed ourselves, namely to use two-body data to fix the t-matrix and knockout reactions to map momentum space wavefunctions is in principle impossible, since the binding of the target nucleon requires that we use an off-shell t-matrix[3] even if we could ignore the distortions which have occupied much of our time here. Now I must try to reverse the flow of the discussion and try to convince you that the program may still be viable from a practical standpoint.

First of all, the unitary transformations discussed above may well provide us with a freedom which we can exploit. If we picture the interaction between the two nucleons as being mediated by the exchange of various mesons, which interact among themselves, then the existence of a nearby nucleon with its attendant mesons could alter the properties of a meson mediating the interaction, and this clearly gives rise to a three-body interaction. Now, the many-body Schroedinger Equation is hard enough to solve for simple two-body interactions and we certainly must hope that higher body interactions may be treated perturbatively.

Here, then, is a possible role for our unitary transformations. One could seek a transformation which eliminates the three-body interaction terms, putting the strength into an effective two-body interaction and, of course, into more-body terms. If these could be kept small and short-ranged, the entire many-body problems could be simplified. In these considerations, we are aided by the Pauli Principle, which restricts the maximum number of nucleons per spatial state to four. Since many-body interactions require the close proximity of the many bodies, this gives a natural cut-off to the problem.

Furthermore, we are not completely ignorant of the properties of the two-body interaction. It has been known since 1933[4] that the force is short ranged (no extremely heavy nuclei in nature), but not zero ranged. This last can be seen from a simple argument. The deuteron has essentially zero binding energy, which implies that the kinetic and potential terms are equal. If we add another nucleon, the larger reduced mass implies less kinetic energy while, assuming only one interaction, the potential is the same. Hence the binding energy is some finite fraction of the kinetic energy. For zero range the kinetic term is infinite, hence so is the binding energy, in contradiction to observation. Further, the fact that the s-wave phase shifts change sign indicates a short-range repulsion. In fact, we have pretty good theoretical reasons for considering the interaction as being given by one-pion exchange at large distances, some form of attraction at distances around one fermi from various two-pion exchanges, and some short-ranged repulsion.

The "realistic" potentials all have these general characteristics, and, when off-shell t-matrices are calculated with them, they do not differ very much until the parameters are considerably away from the energy shell. To elucidate this point, Picker, Redish and I modelled the scattering wavefunction in the following way.[5] We chose one potential (the Reid Soft-Core)[6] to hold beyond 1.43 fermi, fixed the wavefunction at infinity by the phase-shift, integrated the Schroedinger Equation in to 1.43 fermi, and parameterized the interior with a low order polynomial. We then varied that interior wavefunction and calculated t-matrices. Considerable variation of the interior region was required to get noticeable effects near the on-shell point. This even held when we forced a node in the wavefunction at the "hard-core" radius.[7]

The general effect on knock-out reactions may be seen in the work of Redish, Lerner, Haftel and myself,[8] in which we calculated, in plane-wave approximation, the predictions of several potentials. While there was some scatter in the predictions, the differences were dominated by the different fits to on-shell scattering data rather than the off-shell extensions.

Lest I sound too sanguine about the suppression of off-shell uncertainties, let me bring up a worrisome point. The uncertainties in the predicted shapes of angular distributions for knock-out reactions discussed above is less than the modification due to the use of distorted waves. These distorted waves are generated by phenomenological optical potentials wherein a radial form is assumed and parameters are fit to elastic scattering. A consistent discussion of nucleon-nucleon model dependence should require that the optical potential be calculated from the same nucleon-nucleon model as input. The argument against being overly concerned about this point usually is that alterations of the scattering functions for the protons on the target and residual nuclei over the interior of the nucleus will primarily affect the magnitude of the cross-section, not its shape. This is a point which may be investigated by performing unitary transformations restricted to the interior of the nucleon-nucleus system. Such studies should be carried out.

The existence of strong distortions raises another relevant theoretical worry. Throughout our discussions so far we have assumed that the knock-out amplitude can be factorized as the product of a two-body t-matrix and an integral over distorted waves and the bound-state overlap function. However, the distorted waves are smeared in momentum space, so that the correct procedure involves another integral over fully-off-shell t-matrices. A determination first of the accuracy of the approximation and second of the optimum values of the t-matrix parameters requires a comparison with a full calculation, which remains to be done. If the result of such a study would be that the parameters of the relevant t-matrices are much farther off-shell than plane-wave calculations suggest, which I do not expect, we would have to modify the discussions of model dependence above.

Let me summarize these remarks. The absence of a basic theory of strong interactions precludes a discussion of the two-body interaction or t-matrix. However, if we are guided by the prejudices developed over the last few decades, the uncertainties we will tolerate in reasonable t-matrices are small compared to many-body uncertainties. In particular, viewed as a laboratory for studying the off-shell two-body t-matrix, the nucleon knock-out reaction could tell us if our prejudices were grossly wrong, but will not distinguish between models very easily.

What about the use of these reactions to study momentum space wavefunctions, which is the subject of this workshop? There again the prospects are clouded by many-body effects. If, however, one uses this reaction in conjunction with other reactions, such as (e,e'p) to understand, or at least normalize out, reaction questions, and if one chooses geometries in which the momentum dependence of the overlap function is the most rapidly varying in the problem, it is possible to study that function in unique regions of the variable space.

In short, the field is very promising. It promises hard experiments and difficult analysis, but when all that is done, it promises unique information on the behavior of one and two nucleons which are, in some sense, deeply embedded in the nuclear medium.

FOOTNOTES AND REFERENCES

†Work performed under the auspices of the United States Energy Research and Development Administration.

1) M. I. Haftel and F. Tabakin, Nucl. Phys. $\underline{A158}$, 1 (1970); M. I. Haftel, Phys. Rev. $\underline{C3}$, 921 (1971).
2) Any name references not cited in the literature refer to this workshop.
3) E. F. Redish, G. J. Stephenson, Jr. and G. M. Lerner, Phys. Rev. $\underline{C2}$, 1665 (1970).
4) E. Wigner, Phys. Rev. $\underline{43}$, 252 (1933).
5) H. S. Picker, E. F. Redish and G. J. Stephenson, Jr., Phys. Rev. $\underline{C4}$, 287 (1971).
6) R. Reid, Ann. Phys. (NY) $\underline{50}$, 411 (1968).
7) V. G. Neudatchin et al, University of Maryland Technical Report No. 74-040 (1974).
8) G. J. Stephenson, Jr., E. F. Redish, G. M. Lerner and M. I. Haftel, Phys. Rev. $\underline{C6}$, 1559 (1972).

HIGH RESOLUTION (p,2p) STUDIES AT 800 MeV

Robert K. Cole
University of Southern California, Los Angeles, CA 90207

Quasi-free scattering processes such as (p,2p) and more recently (e,ep) have become very useful tools to investigate the binding energies and the momentum distributions of nucleons in nuclei, especially, the more strongly bound nucleons which cannot be reached with pick up reactions. In particular (p,2p) studies at sufficiently high energy show structure in the missing-energy spectrum that is interpreted as arising from the ejection of deeply bound protons. However, much of the experimental work has been with relatively low incident energies ($\lesssim$ 100 MeV) and, consequently, has investigated primarily the lighter nuclei and the lesser bound protons of the medium to heavy nuclei. The mean free path of protons at these energies is relatively short, comparable to or less than the radius of the nucleus. Thus, there is a relatively high probability that a proton may interact with more than one nucleon in the nucleus. Consequently, most observed (p,2p) events involving only one interaction between the incident particle and a target proton occur near the nuclear surface which is dominated by the upper shell nucleons. The discrete binding energy spectrum is observed superimposed on a background produced by multi-scattering processes. The angular correlations between the two protons are also distorted by multiple collisions and theoretical interpretation requires distorted wave calculations. The higher energy experiments should suffer less from multiple nucleon scattering but experimental difficulties have resulted in much poorer energy resolution; typically about 5 MeV, and smaller statistical accuracy. This energy resolution is adequate to resolve the shell structure of only the light nuclei.

The additional experimental difficulties of the higher energy (p-2p) studies are counterbalanced by a more straight-forward interpretation of the experimental data. Multiple scattering is less and the assumption that the (p,2p) event represents an interaction between the incident proton and one target proton while the other nucleons are spectators is probably closer to reality.

Both the plane wave Born approximation, PWBA, and the distorted wave Born approximations DWBA, are factored into two recognizable terms. The first term represents the proton-proton interaction and is often replaced by the free (p-p) cross sections at an appropriate energy and angle. The second factor is recognized as the Fourier transform of the bound state wave function with respect to the momentum transferred to the residual nucleus and called the

distorted momentum distribution for DWBA. Multiple scattering of
the protons is taken into consideration by means of distorted
waves. However, the optical potentials that give the best fit
to the (p,2p) data may not be, in general, identical to those
obtained from elastic scattering since they represent different
interactions. The higher energy experiments reported qualitatively
support the conclusions of PWBA although the cross section is
reduced because of absorption and the maxima and minima are
shifted a little. Also, the plane wave and distorted momentum
distribution have similar shape and width.

However, the hole state produced is not necessarily an eigen
state of the residual nucleus. One may expect that the resulting
energy spectra would be complex splitting into many levels. So
far, structure has only been observed for the protons knocked out
of the outer shells. This is explained by the very short lifetime of a hole state in an inner shell and results in a broad
width. Recently there has been discussion concerning the time
it takes the nucleus to rearrange. If rearrangement time is comparable to the escape time of the two protons, then an energy
dependence of the shape of the energy spectra will result.

Most of the high energy experiments have obtained an energy
resolution of several MeV and have concentrated on the lighter
elements with identification of 1s, 1p, 1d and 2s states. However,
a number of questions remain open. These include the identification
of the deeply bound states in nuclei with A greater than about 40,
the structure of these strongly bound states, and the resolution
of the structure of the upper shell states especially for heavier
elements. Another interesting problem is the validity of the
factorization of the theoretical cross section in particular for
the deeply bound protons.

We plan to do high energy (p,2p) studies (800 MeV incident
protons) with quite good energy resolution, summed energy spectra
with an overall energy resolution of about 500 keV. Because of
the large kinematic effect at these energies, this also requires
very good angular resolution for both detectors. Since only one
high resolution magnetic spectrometer, HRS, will be available and
the possibility of another is quite remote, we developed an
alternative detector. The detector consists of 8 - one cm thick
intrinsic Ge diodes which will stop 200 MeV protons, 270 MeV
deuterons and 330 MeV tritons. Although the energy resolution
that we can expect to obtain from the Ge telescope is somewhat inferior to that obtainable with high resolution magnetic spectrometers, it is adequate to do these experiments. On the other hand
the Ge detector can obtain at one time the complete energy spectra
of all charged particles that are stopped. This is an important
advantage compared to magnetic spectrometers for correlation ex-

periments in particular.

However, the question of obtaining good energy resolution is not the most serious problem for stopping detectors in this energy range. Major problems are associated with the relatively large probability that an incident particle will either scatter out of the detector or will undergo a nuclear reaction in the detector material before it reaches the end of its range. If either occurs, the resulting signal generated by the detector will be anomalously small and could be interpreted as a lower energy particle or a particle of a different type. For example, a 200 MeV proton entering on the axis of our Ge telescope (8 cm in length by 3 cm diameter) has a 10% probability that it will scatter out of the telescope and about 25% probability of initiating a nuclear reaction before losing all its energy. Thus, the pulse height spectrum produced by a mono-energetic beam of particles consists of a sharp peak at full energy plus a low energy tail. This response function is particularly troublesome for measurements of reactions other than elastic scattering where the events of interest are superimposed on the tail produced by the higher energy particles which may be greater in number.

Our solution to this problem is to divide the stopping detector into a number of elements or laminae and to measure the energy deposited in each element. A comparison of the measured energy loss in each element to that predicted for a particle of incident energy equal to the energy measured by this detector will reveal if a nuclear interaction or scattering out has occurred.

The detector telescope consists of eight diodes each 1 cm thick by 3 cm diameter and was fabricated for us by the Berkeley detector group from p-type intrinsic germanium. A rectifying junction was made by evaporating a thin film of Li on one surface and annealing for about 5 minutes at 200° C. This formed an ~ 15 µm dead layer which is sufficiently small so that Landau energy loss fluctuations do not impair the energy resolution of the instrument. An ohmic contact on the other surface is formed by the evaporation of a thin Pd film.

Each diode is mounted on its individual holder which fits into a machined slot in a copper block. The separation between adjacent diodes is 5 mm. The copper mounting block is suspended from the lid of the detector vacuum chamber by stainless steel tubes through which liquid nitrogen refrigerant may circulate. The removeable lid also contains the electrical feed-throughs and seals the top of the detector vacuum chamber. Particles enter the cylindrical vacuum envelope through a thin window.

We have recently used this detector to study clustering in ^{6}Li and ^{7}Li with the 800 MeV external proton beam at LAMPF. The principal reactions studied were ^{6}Li(p,pd) ^{4}He and ^{7}Li(p,pt) ^{4}He.

The protons were detected and momentum analyzed by a magnetic spectrometer of momentum resolution $\sim 1\%$. The conjugate particles were measured by a detector system which included two plastic scintillation counters for coincidence timing and maximum solid angle definition; a small helical wire proportional chamber which determined the vertical and horizontal position of each particle; a 1 mm thick Si surface barrier diode for dE/dx measurements of particles that stopped in the first Ge diode; the 8 cm Ge telescope; and a final plastic scintillator to veto particles that pass through the entire system.

The performance of the Ge diode telescope was evaluated by elastic scattering 800 MeV protons from CH_2 and CD_2 targets and detecting both coincident particles. These reactions produced protons and deuterons of well defined energies incident on the detector. Detailed results will be published later, but, in the brief, we have now developed a computer program to analyze each event and reject those that have had an inelastic scattering in the detector material.

Preliminary results of the $^6Li(p,pd)$ ^4He study were presented at the April 1976 Washington D.C. APS meeting. In this preliminary analysis we have simply summed the energies of the coincidence proton and deuteron over the total angular acceptance of both detectors. The resulting energy resolution was determined by the momentum resolution of the magnetic spectrometer over its full angular width and was more than adequate to separate the (p,pd) ground state events from other possible reactions. The energy resolution of the Ge telescope could not be obtained even from the elastic $^2H(p,pd)$ measurements since it was hidden by the energy spread of the incident beam (1 to 2 MeV) and our minimum angular resolution of 3 m radian for the detectors. Results of the $^6Li(p,pd)$ ^4He at $\theta_p = 30°$ and $\theta_d = -62°$ are shown in fig. 1.

For our (p,2p) studies we propose to first measure binding energy spectra with an overall resolution of ~ 500 keV. The measurements will be coplanar and asymmetric with the HRS detecting the higher energy proton ~ 600 MeV and the Ge telescope system measuring the lower energy coincident protons. Targets of interest for initial studies include ^{26}Mg, ^{27}Al, ^{28}Si and ^{40}Ca as representatives of the 1d-2s shell nuclei and ^{59}Co, ^{58}Ni, and ^{60}Ni with 1f-2p shell nucleons.

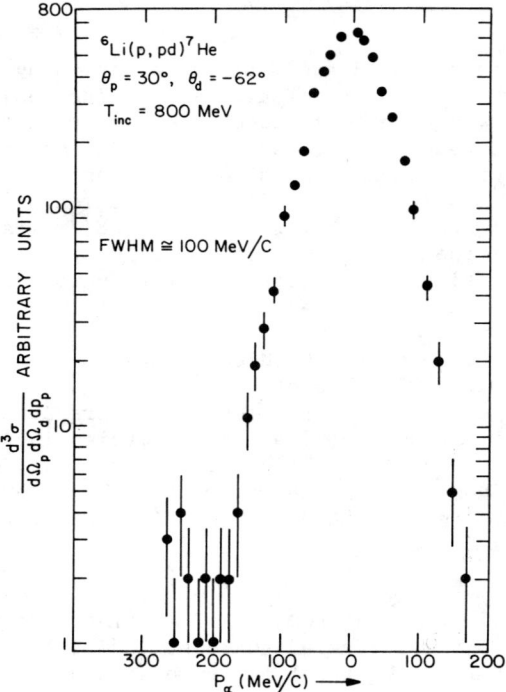

Figure 1. Recoil momentum distribution for ^{6}Li(p,pd)^{4}He at E_p = 800 MeV.

QUASI-FREE $(\pi,\pi N)$ SCATTERING [†]

V.E.Herscovitz, Th.A.J.Maris, P.M.Mors and C.Schneider
Instituto de Física, UFRGS, 90000 Porto Alegre, RS, Brasil

ABSTRACT

Quasi-free $(\pi,\pi N)$ scattering in the resonance region is discussed.

Nuclear Momentum Space Wave Functions in individual nuclear shells can be investigated by quasi-free $(p,2p)$[1] and $(e,e'p)$[2] scattering. In this note we shall discuss some essential features of quasi-free $(\pi,\pi N)$ scattering.

For pion energies far over the resonance energy, i.e. for beam energies higher than 300 MeV, the mean free path of a pion in nuclear matter is of the order of several fermis. In this energy region quasi-free $(\pi,\pi N)$ scattering in nuclei is expected to be describable by the same type of Distorted Wave Impulse Approximation[1] as applies to quasi-free $(p,2p)$ and $(e,e'p)$ processes. Also the nuclear information on momentum distributions obtainable at these energies should be comparable to the ones from $(p,2p)$ and $(e,e'p)$ experiments.

For pion energies in the resonance region, i.e. for $150 < E_\pi < 250$ MeV there are however some important differences as compared to medium energy quasi-free $(p,2p)$ scattering:

a) Because of the large free resonance cross section, the mean free path of the incoming pion is in the centre of the nucleus less than one fermi, which is about one quarter of the mean free path of a medium energy proton and of the radius of a light nucleus.

b) The de Broglie-wavelengths of the incoming and emerging pions are of the order of the radius of a light nucleus and at least twice the wavelength of a medium energy proton.

c) Because of the resonance, the free scattering matrix elements are fast varying with energy and momentum, compared to the slowly varying medium energy proton-proton cross section.

As a result of points a) and b), the usual WKB-approximation for the wave distortions, simulating the initial and final state interactions, is suspect. One might think that this could only mean a calculative complication, demanding a phaseshift analysis for the in and outgoing waves, but this conclusion would be incorrect.

[†] Work partially supported by CNPq (Brasil), FINEP (Brasil) and GTZ (Germany).

The main reason is that a breakdown of the WKB-approximation also prohibits the factorization of the quasi-free correlation cross section into an, in general modified, "free" (π-N) cross section and a distorted momentum distribution. This is already clear from the fact that the "free" cross section in this product is taken for one set of momenta, deduced from the selected kinematics of the process, which would be unrealistic if the initial and final state momenta would introduce such a large momentum spread as to invalidate the WKB approximation. That is in particular true because of the strong momentum dependence of the resonance cross section. We remark that a non-factorizability of the cross section not only increases the computational task by at least an order of magnitude over the one of a factorized phase shift calculation (which is again considerably longer than the WKB-one), but also spoils the direct interpretation[3,4] of the experimental results of quasi-free scattering.

We have calculated the distortion by the initial and final state interactions for quasi-free coplanar (π^+,π^+p) scattering of 200 MeV incident pions knocking out 2s-state protons of ^{40}Ca and have made contour plots showing the relative contributions of the various regions of the nucleus to the quasi-free cross section. The distortion was computed with two types of calculations:

a) With the WKB method, using various distorting potentials.

b) With a partial wave analysis for the incoming and outgoing waves.

The conclusions of this investigation in respect to $(\pi,\pi N)$ scattering in the resonance region are:

a) If one neglects the real optical pion potentials, the WKB is a good approximation.

b) With the inclusion of the real pion potentials as following from multiple scattering theory, in particular the outgoing pion shows strong focusing effects[5] which makes the WKB and the factorization in general bad approximations.

c) In very special geometries the effect mentioned under b) may be minimized and the reaction might then be useful for an investigation of the behaviour of the resonance in the nuclear surface.

A more detailed account of this work will be published.

REFERENCES

1. Th.A.J.Maris, P.Hillman and H.Tyrén, Nucl. Phys. $\underline{7}$ (1958) 1.
2. G.Jacob and Th.A.J.Maris, Nucl. Phys. $\underline{31}$ (1962) 139.
3. J.Hüfner, H.J.Pirner and M.Thies, Phys. Lett. $\underline{59B}$ (1975) 215.
4. L.C.Liu and P.Huguenin, Helv. Phys. Acta. $\underline{46}$ (1973) 201.
   L.C.Liu, Nucl. Phys. $\underline{A223}$ (1974) 523.
5. I.E.McCarthy, Nucl. Phys. $\underline{11}$ (1959) 574.

A KNOCKOUT REACTION STUDY WITH A POLARIZED PROTON BEAM*

P. Kitching, C.A. Miller, D.A. Hutcheon,† A.N. James,‡
W.J. McDonald, J.M. Cameron, W.C. Olsen and G. Roy
University of Alberta, Edmonton, Alberta, Canada

ABSTRACT

The knockout reaction $^{16}O(\vec{p},2p)$ has been studied with a polarized beam at 200 MeV. Strong j-dependence of cross section is observed, indicating that under appropriate kinematic conditions knockout protons may be polarized.

INTRODUCTION

I wish to report on one of the first experiments done with the polarized beam at TRIUMF. For those who may not be familiar with the TRIUMF machine, it is a cyclotron with two extracted proton beams, readily variable over the energy range 200-500 MeV. Since March we have had available polarized beams of intensity 20 to 30 nA at 75% polarization. A group from the University of Alberta is involved in a series of quasi-elastic scattering experiments—$^{12}C(p,pd)$, $^{12}C(p,p\alpha)$, $^{12}C(p,pn)$—but I will talk here only of our first results from a study of proton knockout in ^{16}O using a polarized beam.

The purpose of the $^{16}O(\vec{p},2p)$ study was to investigate a proposal by Jacob et al.[1] that one should be able to polarize protons in the nucleus, in the sense that one could choose kinematic conditions to favour knockout of protons with a particular spin orientation. The proton knockout is considered in the impulse approximation, meaning roughly that interaction of the projectile and struck proton takes place as though it were free p-p scattering unaffected by the presence of the residual nucleus; the momentum of the knocked-out proton just before the interaction should be equal and opposite that of the residual nucleus. For beam energies available at TRIUMF the impulse approximation is expected to be good.

The possibility of polarizing knockout protons comes about due to (1) the mean free path of protons in nuclear matter increasing significantly as their energy increases from 50 to 200 MeV, and (2) the p-p scattering cross section at intermediate energies being much larger for parallel proton spins than for anti-parallel spins. The energy dependence of mean free paths means that when one of the outgoing protons has much lower energy than the other there is a tendency for the reaction to take place near the surface of the nucleus on the side which minimizes absorption of the lower-energy proton. This is illustrated in Fig. 1 where, if 1 is the high- and 2 the low-energy proton, case (b) will be favoured over (a), i.e. the reaction occurs on the side of the nucleus in the direction $\vec{k}_2$.

---

*Work supported in part by the Atomic Energy Control Board and by the National Research Council of Canada.
†Present address: TRIUMF, Vancouver, B.C., Canada V6T 1W5.
‡On leave from University of Liverpool, England.

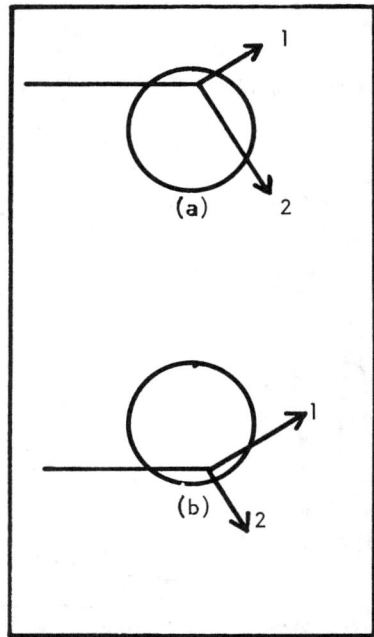 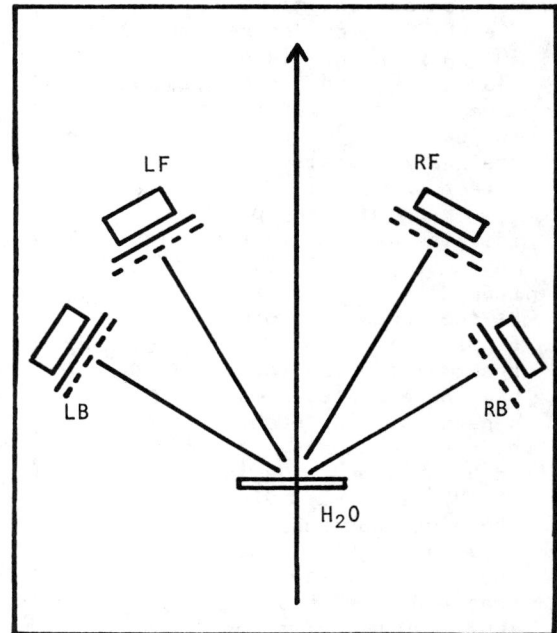

Fig. 1. Absorption of low-energy proton 2 favours case (b) in (p,2p) reaction.

Fig. 2. Schematic view of target and detector telescope arrangement.

Detection angles and energies of protons 1 and 2 determine the momentum of the residual nucleus and hence, in the impulse approximation, that of the struck proton $\vec{k}_3$, which with $\vec{r}_3$ determines its orbital angular momentum $\vec{\ell}_3$. Nuclear spin-orbit coupling makes the intrinsic spin of the proton either parallel or anti-parallel to ℓ, and so by choosing protons from one or the other of $j=\ell-1/2$ or $j=\ell+1/2$ subshells it should be possible to select spin-up or spin-down protons in the nucleus. If this is possible, then the large difference between ↑↑ and ↑↓ cross sections should result in $(\vec{p},2p)$ asymmetries which are different for different subshells.

EXPERIMENTAL METHOD AND RESULTS

The experimental arrangement is shown in Fig. 2. Four counter telescopes, each consisting of a multiwire proportional counter, thin plastic scintillator and 3 in. × 5 in. φ NaI crystal, were set at angles 30° and 58° on the left and right sides of the beam. The target was 1 mm thick H_2O and the beam energy 200 MeV. The combinations LF·RB or RF·LB correspond to angles of p-p scattering from the hydrogen in the water target and were used to give normalization of $(\vec{p},2p)$ cross sections and the beam polarization in terms of the free p-p scattering differential cross section and analyzing power. The LF·RF coincidences (30°-30°) give $\vec{k}_2$ nearly perpendicular to $\vec{k}_3$, and the effect is expected to be large, while for LB·RB (58°-58°) $\vec{k}_3$ and $\vec{k}_2$ are nearly parallel and the effect should be small.

From measured proton energies we compute the 'missing energy' $E_0-E_1-E_2$ which should be the binding energy of the knockout proton. Figure 3 is a 'missing-energy' spectrum showing the $p_{1/2}$ and $p_{3/2}$ hole states cleanly separated. For each of these groups the counts were added in bins 10 MeV wide in E_1 (20 MeV in E_1-E_2) yielding the distributions shown in Figs. 4 and 5. Cross sections have been adjusted to values that would be obtained for 100% beam polarization. The points labelled spin up (down) include both the data for $E_L > E_R$ with spin up (down) and $E_L < E_R$ with spin down (up) which are related by a 180°

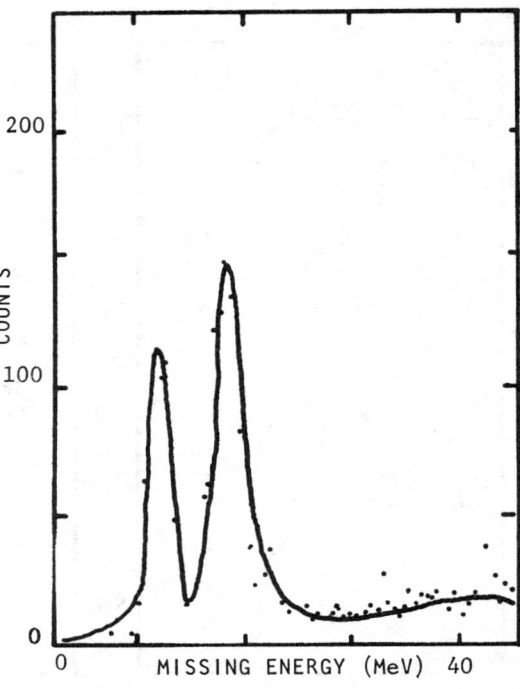

Fig. 3. Missing-energy spectrum showing peaks for $p_{1/2}$ and $p_{3/2}$ hole states.

rotation about the beam axis. The 30°-30° curves show large polarization effects of the sort expected and of opposite sign for the $p_{1/2}$ and $p_{3/2}$ states, while for 58°-58° no polarization effect is seen, again as expected.

The curves in Figs. 4 and 5 are calculated by a distorted wave impulse approximation program written by C.A. Miller. This was a partial wave calculation in which the optical potential was of Woods-Saxon form including surface absorption, and the bound-state wave function was calculated with a binding potential of Woods-Saxon shape with a spin-orbit term and Coulomb repulsion, the geometrical parameters being chosen to reproduce high-energy (e,e) and (p,2p) data. Spin-orbit distortion effects of the optical potential were neglected. It is seen that the model reproduces very well the observed asymmetries and cross sections with the exception of the $p_{3/2}$ cross sections at 58°-58°. A similar anomaly was seen in earlier ^{16}O(p,2p) work at 460 MeV.[2]

It has been suggested by D.F. Jackson[3] that inclusion of spin-orbit distortion terms, which can alter the polarization state of the incident protons, could have a significant effect on $(\vec{p},2p)$ asymmetries and even on cross sections. Our asymmetry results show no sign of such spin-orbit distortion, but these data cover a very restricted range, of course.

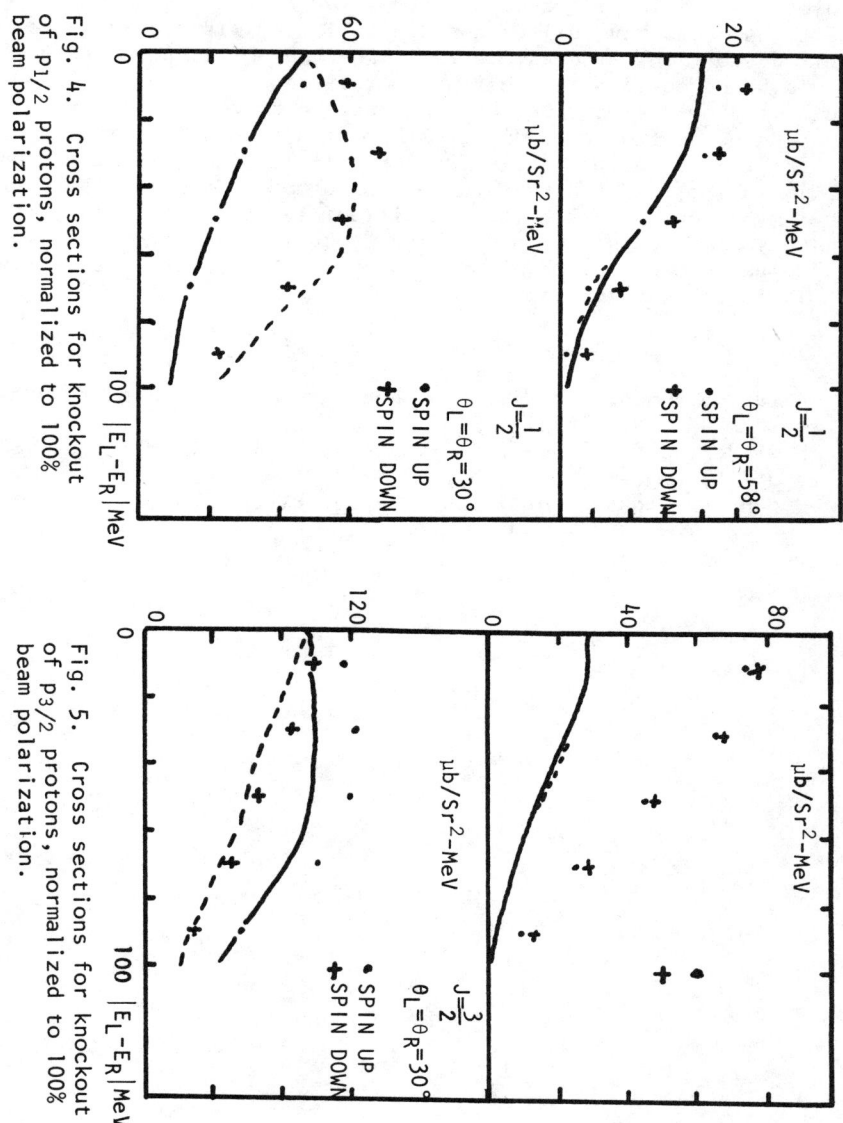

Fig. 4. Cross sections for knockout of $p_{1/2}$ protons, normalized to 100% beam polarization.

Fig. 5. Cross sections for knockout of $p_{3/2}$ protons, normalized to 100% beam polarization.

FUTURE MEASUREMENTS

Measurements of $^{16}O(\vec{p},2p)$ at 200 MeV over a much wider range of angles are now in progress. We also plan to study $^{40}Ca(\vec{p},2p)$. These additional data may tell us what role, if any, is played by spin-orbit distortions; the asymmetry in knockout of 2s protons from ^{40}Ca should be especially interesting in this regard since the DWIA predicts it should simply be that of free p-p scattering.

There are two quite different ways that one might hope to use the ability of $(\vec{p},2p)$ to distinguish $j=l-1/2$ from $j=l+1/2$ protons. One is in looking at structures of the broad deep-hole states, the other is in nuclear structure studies of individual levels. In the coming year we expect to commission a magnetic spectrometer which will give an order of magnitude improvement in energy resolution available for such studies.

REFERENCES

1. G. Jacob, Th.A.J. Maris, C. Schneider, and M.R. Teodoro, Phys. Lett. B45, 181 (1973).
2. H. Tyren, S. Kullander, O. Sundberg, R. Ramachandran, P. Isacsson, and T. Berggren, Nucl. Phys. 79, 321 (1966).
3. Daphne F. Jackson, Nucl. Phys. A259, 221 (1976).

THREE-BODY COLLISIONS INVOLVING BREAKUP

Ian H. Sloan
School of Mathematics, University of New South Wales,
Sydney, N.S.W. 2033, Australia.

ABSTRACT

This paper gives a non-technical introduction to breakup processes caused by a particle incident on a bound pair. A brief discussion of the theory is followed by a review of exact calculations and approximation methods. The paper concludes with a short discussion of the special problems arising when the particles are charged.

INTRODUCTION

The aim of this talk is to provide a brief overview of three-body breakup collisions, for the benefit of non-specialists.

Let us consider particle 1 incident on a bound state of particles 2 and 3, producing a final state with all three particles free. The particles might be three nucleons, or two electrons and an atomic core, or two nucleons and a nuclear core. However, the particles will be assumed to be structureless, and, to simplify the discussion, spin and isospin will be neglected.

For the present I shall assume that all forces are short-range two-body forces. However, towards the end I shall say a little about the Coulomb case, a case with a rather different flavour in both theory and practice.

I shall begin with an informal discussion of the theory for the short-range case, and then say something about applications and methods of calculation.

THEORY OF THE BREAKUP PROCESS

A convenient starting point is the following expression[1] for the amplitude f for the process $1 + (2,3) \to 1 + 2 + 3$:

$$f(\vec{p}_1',\vec{p}_2',\vec{p}_3'; \vec{p}_1) = (\Phi_{\vec{p}_1'\vec{p}_2'\vec{p}_3'}, (V_1 + V_2 + V_3)\Psi^+_{\vec{p}_1}), \qquad (1)$$

where $V_1 = V_{23}$ is the potential for particles 2 and 3; $\vec{p}_1$ is the initial c.m. momentum of particle 1; $\vec{p}_1',\vec{p}_2',\vec{p}_3'$ are the final c.m. momenta of particles 1,2,3 (only two of these being independent, since $\vec{p}_1' + \vec{p}_2' + \vec{p}_3' = 0$); $\Phi_{\vec{p}_1'\vec{p}_2'\vec{p}_3'}$ is the final plane wave state; and $\Psi^+_{\vec{p}_1}$ is the appropriate exact wave function, with asymptotic form corresponding to particle 1 incident, plus outgoing scattered waves for all particles.

The exact wave function can be written explicitly as

$$\psi^+_{\vec{p}_1} = [1 + G(V_2 + V_3)]\Phi_{\vec{p}_1}, \qquad (2)$$

where $\Phi_{\vec{p}_1}$ is the initial unperturbed wave function (a plane wave multiplied by the bound state wave function of 2 and 3); G is the appropriate Green's function at energy E, i.e.

$$G = (E + i\varepsilon - H_0 - V_1 - V_2 - V_3)^{-1};$$

H_0 is the three-particle kinetic-energy operator; and $\varepsilon \to 0+$.

From (1) and (2), the amplitude can be written as

$$f(\vec{p}_1',\vec{p}_2',\vec{p}_3'; \vec{p}_1) = (\Phi_{\vec{p}_1'\vec{p}_2'\vec{p}_3'}, U_{01} \Phi_{\vec{p}_1}), \qquad (3)$$

where the transition operator U_{01} is defined by

$$U_{01} = V_1 + V_2 + V_3 + (V_1 + V_2 + V_3) G(V_2 + V_3). \qquad (4)$$

It can easily be shown that the first term V_1 does not contribute to the physical amplitude (3), so from now on we shall leave it out.

To find a suitable integral equation for U_{01}, let us proceed in the following non-rigorous fashion: First, in terms of the free-particle Green's function,

$$G_0 = (E + i\varepsilon - H_0)^{-1}, \qquad (5)$$

the full Green's function satisfies the identity

$$G = G_0 + G_0(V_1 + V_2 + V_3)G,$$

leading to the series

$$G = G_0 + G_0(\Sigma V_i)G_0 + G_0(\Sigma V_i)G_0(\Sigma V_i) G_0 + \dots .$$

On substitution into (4) we obtain the Born series for U_{01},

$$U_{01} = V_2 + V_3 + V_1 G_0 V_2 + V_2 G_0 V_2 + V_3 G_0 V_2 + \dots . \qquad (6)$$

The sum contains all terms of the form $V_i G_0 V_j G_0 \dots G_0 V_n$, where V_n on the right is V_2 or V_3.

We now sum the subsequences which involve only the potential V_i, by defining

$$T_i = V_i + V_i G_0 V_i + V_i G_0 V_i G_0 V_i + \dots . \qquad (7)$$

This is the Born series for the Lippmann-Schwinger equation

$$T_i = V_i + V_i G_0 T_i, \qquad (8)$$

hence T_i is essentially the t-matrix for the off-shell scattering of the pair i in the presence of the third particle. [The third particle enters through its contribution to the kinetic-energy term in (5).]

Then (6) becomes

$$U_{01} = T_2 + T_3 + T_1 G_0 T_2 + T_3 G_0 T_2 + \ldots . \qquad (9)$$

The sum here includes all terms of the form $T_i\, G_0\, T_j\, G_0 \ldots G_0\, T_n$ where T_n is T_2 or T_3, with the restriction that $i \neq j \neq \ldots \neq n$; i.e. in contrast to (6), *the same pair never appears twice in succession*.

Equation (9) is a "multiple-scattering" series for U_{01}. Each term in this series corresponds to a particular sequence of successive scatterings of different pairs of particles, beginning on the right with either the pair (13) or the pair (12), because particle 1 is the incident particle.

It can be shown rigorously from (4) that U_{01} can be written as the sum of three terms,

$$U_{01} = X_1 + X_2 + X_3,$$

which satisfy the Faddeev equations,[2]

$$\begin{aligned} X_1 &= T_1\, G_0(X_2 + X_3) \\ X_2 &= T_2 + T_2\, G_0(X_3 + X_1) \\ X_3 &= T_3 + T_3\, G_0(X_1 + X_2). \end{aligned} \qquad (10)$$

The iterative solution of these equations is just the multiple-scattering series (9), with X_i being the sum of all those terms that have T_i on the left.

The Faddeev equations are mathematically well behaved,[2] essentially because of the absence of diagonal terms in the kernel; for instance, the first of Eqs.(10), i.e. the equation for X_1, does not have X_1 on the right hand side. In contrast, the corresponding Lippmann-Schwinger equation, with potentials rather than t-matrices in the kernel, has diagonal terms in the kernel, and these involve only a single potential V_i, so that the kernel fails to be square-integrable even if it is iterated any number of times. Thus there are good mathematical as well as physical reasons for the occurrence of t-matrices rather than potentials.

The Faddeev equations can either be used to suggest approximation methods, or be attacked directly. The second kind of application is more soundly based than the first, but so far has apparently only been carried out for the three-nucleon system.

EXACT THREE-NUCLEON CALCULATIONS

Exact breakup calculations for the three-nucleon system were first carried out in 1966, for the special case of so-called separable potentials.[3] More recently, in an impressive calculation by Kloet and Tjon,[4] the case of S-wave spin dependent *local* potentials has been treated exactly. I shall say a few words about the local potential case.

What is involved in such a calculation? The Faddeev equations in the momentum representation are integral equations in two vector variables (in the c.m. system) say $\vec{p}_1'$ and $\vec{q}_{23}' = (m_3\vec{p}_2' - m_2\vec{p}_3')/(m_2 + m_3)$. After carrying out partial-wave analysis over the angles of the vectors, one finishes with coupled, two-dimensional integral equations. Kloet and Tjon have solved these equations numerically for a sufficient number of three-body partial waves, by an ingenious use of Padé approximant techniques for summing the multiple-scattering series.

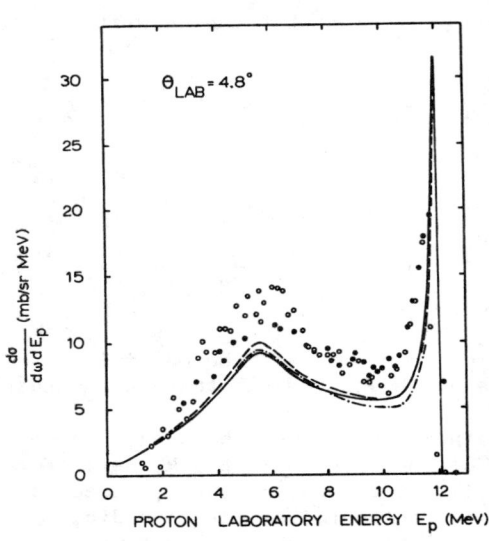

Fig.1. Proton spectra at proton lab angle of $4.8°$, from Ref.4.

To indicate what can be achieved, Figs.1-3 show Kloet and Tjon's proton spectra at various angles, for 14.4 MeV neutrons incident on deuterons. The solid and dashed curves correspond to two slightly different choices[4] of local potential; the dash-dotted curve corresponds to a separable potential. The experimental results shown by open circles[5] are probably preferable to the older results[6] shown by the closed circles.

How far can exact calculations of this kind be extended? The number of coupled equations with which we are faced depends steeply on the number and multiplicity of the two-body partial waves needed to describe the two-body interaction. In this respect the three-nucleon problem at fairly low energies is a fortunate one, in that the two-body S-waves are dominant. The problem becomes considerably harder if one of the particles is a nuclear core, as in the general (p,2p) reaction, since it is then no longer reasonable to assume that the two-body interactions are purely S-wave. It may also be remarked that the inclusion of non-central forces in the three-nucleon problem makes that problem very much harder. Finally, the calculations become rapidly more difficult as the energy is increased, mainly because of the resulting increase in the number of important two-body partial waves.

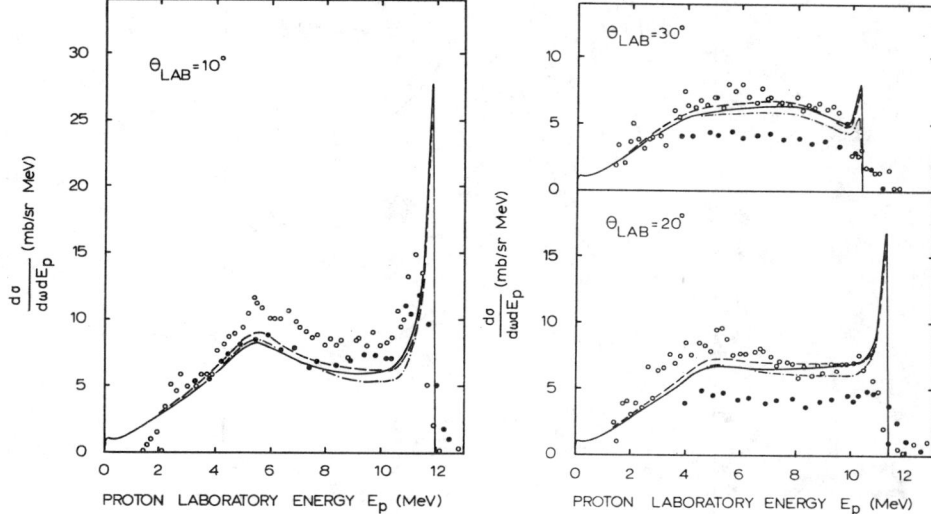

Fig.2. Proton spectra at proton lab angle of 10°, from Ref.4.

Fig.3. Proton spectra at proton lab angles of 20° and 30°, from Ref.4.

APPROXIMATE CALCULATIONS

Wherever exact calculations are not feasible, we must turn to approximation methods. Let us consider first the impulse approximation,

$$U_{01} \simeq T_2 + T_3,$$

obtained by retaining only the first-order terms in the multiple-scattering series (9). [This approximation is not quite the original impulse approximation proposed by Chew,[7] in that Chew's final state includes a scattering wave function for the relative motion of the pair (23), instead of, as here, a plane wave.]

The impulse approximation, and similar higher-order approximations from the multiple-scattering series, are attractive in that each term has a clear physical interpretation. Nevertheless, such approximations are not necessarily reliable computational tools, except at very high energies. As an example, consider the impulse approximation and higher-order multiple-scattering approximations for the case considered above, of the proton spectrum from n-d breakup at 14.4 MeV. [The two-body potentials used in obtaining the following results[8] were separable rather than local, but the difference seems to be not significant.]

The results[8] are shown in Fig.4, with the impulse approximation labelled by 1, the second- and third-order multiple-scattering approximations by 2 and 3, and the exact results by E. The first

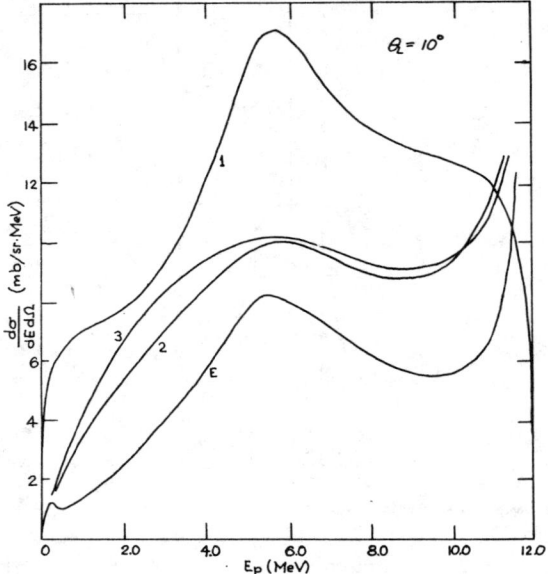

Fig.4. Proton spectra[8] at proton lab angle of $10°$ from 1st, 2nd and 3rd orders of multiple-scattering (reduced on the figure by factors of 2, 7 and 15 respectively), compared with the exact solution (E) for the same potential.

three curves have been reduced by factors of 2, 7 and 15 respectively to bring them into scale, so that contrary to appearances, the multiple-scattering series is actually diverging rapidly. Indeed, for this model it is known[9] that the multiple-scattering series actually diverges at all energies below 50 MeV, and that even at 100 MeV the convergence is very slow.

If one of the bodies in our three-body system is actually a nuclear core, the intuitive merit of the impulse approximation in the above form becomes much weaker, because of the strong distortion produced by the core, and some form of distorted-wave approximation, e.g. a distorted-wave Born approximation,[10] or distorted-wave t-matrix approximation,[11] would seem almost essential.

Actually, a distorted-wave treatment is indicated even in the three-nucleon system, if the final relative energy of two of the particles, say particles 1 and 2, is small. This is because final-state interaction effects of particles 1 and 2 then dominate the observed cross section. An appropriate starting point for approximations in this case, replacing Eq.(1), is the alternative exact expression[1]

$$f(\vec{p}_1',\vec{p}_2',\vec{p}_3';\vec{p}_1) = (\phi_{\vec{p}_3}\chi^-_{12}, (V_1 + V_2)\psi^+_{\vec{p}_1}),$$

where $\phi_{\vec{p}_3}$ is a plane wave, and χ_{12}^- is the appropriate scattering wave function for the relative motion of particles 1 and 2. Various approximations, including Chew's original impulse approximation,[7] can now be obtained by suitably approximating $\psi_{\vec{p}_1}^+$.

It may be remarked that the region of applicability of distorted-wave methods, as indeed of most approximations in collision theory, is poorly defined. Perhaps the time is approaching when collision theorists will be expected to provide not only formal derivations and plausibility arguments, but also error estimates!

[Note added: An animated discussion took place at this point, with support for the suggestion of error estimates coming from both experimentalists and theorists. However, the point was made in the discussion that in some areas, such as many-body effects in nuclei, useful error estimates could certainly not be given, because of large uncertainties in our understanding of the physics. There is no such excuse, however, when *mathematical* approximations are made. In particular, much theoretical work on few-body collisions is concerned with well-defined approximations within a particular model (say the few-body Schrödinger equation, with perhaps some uncertainty in the potential). In such situations it might well be possible to find useful bounds on the errors arising from the approximations and uncertainties, and so perhaps greatly increase the physical relevance of the calculations.]

THE COULOMB CASE

If the particles are charged, there are a number of difficulties, both theoretical and practical, because of the long range of the Coulomb potential.

On the theoretical side, we have it on the authority of Faddeev[12] that there is as yet no satisfactory integral equation formulation of the three-charged-particle problem at positive energies.

If one is nevertheless tempted to use just the ordinary Faddeev equations in the momentum representation, and simply hope for the best, it may be useful to reflect that one must expect to face, at the very least, the same sort of problems that arise in the corresponding *two*-body problem. In the two-body case, the off-shell Coulomb t-matrix $t(\vec{q},\vec{q}'; E + i\varepsilon)$ exists, and can indeed be constructed explicitly,[13,14] but the limit as q or q' approaches the on-shell value does *not* exist;[14] in fact, it is quite difficult to extract physical scattering information from the off-shell Coulomb t-matrix. The fundamental problem, of course, is that the asymptotic form of the Coulomb scattering wave function is different from the usual form for short-range forces, because of the logarithmically increasing phase term produced by the Coulomb potential. It is therefore only natural that a different definition of the scattering amplitude is required for Coulomb scattering.

The situation is similar, only very much worse, for the three-charged-particle case at postiive energies, and in fact little progress has been made with the kind of momentum-space approach which has been so successful for short-range potentials.

In configuration space, on the other hand, the asymptotic form of the wave function and related questions have been studied extensively, particularly by Peterkop, Seaton and Rudge, in the context of electron-atom ionizing collisions, and this work deserves to be better known. A useful review, covering also approximation methods (mainly variants of the Born approximation) is that of Rudge.[15]

REFERENCES

1. M.L. Goldberger and K.M. Watson, Collision Theory (Wiley, N.Y., 1964), Ch.5.
2. L.D. Faddeev, Soviet Phys. — JETP $\underline{12}$, 1014 (1961).
3. R. Aaron and R.D. Amado, Phys. Rev. $\underline{150}$, 857 (1966).
4. W.M. Kloet and J.A. Tjon, Nucl. Phys. $\underline{A210}$, 380 (1973).
5. N. Koori, J. Phys. Soc. Jap. $\underline{32}$, 306 (1972).
6. K. Ilakovac et al., Phys. Rev. $\underline{124}$, 1923 (1961); M. Cerineo et al., Phys. Rev. $\underline{B133}$, 948 (1964).
7. G.F. Chew, Phys. Rev. $\underline{80}$, 196 (1950).
8. R.T. Cahill and I.H. Sloan, Nucl. Phys. $\underline{A194}$, 589 (1972).
9. I.H. Sloan, Phys. Rev. $\underline{185}$, 1361 (1969).
10. N. Austern, Selected Topics in Nuclear Theory (International Atomic Energy Agency, Vienna, 1963), p.39; W. Tobocman, Theory of Direct Nuclear Reactions (Oxford University Press, London, 1961).
11. L.R. Dodd and K.R. Greider, Phys. Rev. $\underline{146}$, 675 (1966).
12. L.D. Faddeev, in Three Body Problem in Nuclear and Particle Physics, ed. by J.S.C. McKee and P.M. Rolph (North-Holland, Amsterdam, 1970), p.154.
13. J. Schwinger, J. Math. Phys. $\underline{5}$, 1606 (1964).
14. A.C. Chen and J.C.Y. Chen, Phys. Rev. $\underline{A4}$, 2226 (1971).
15. M.R.H. Rudge, Rev. Mod. Phys. $\underline{40}$, 564 (1968).

ON THE (e,2e) REACTIONS IN SOLIDS

N. R. Avery

University of Melbourne, Parkville, Victoria, 3052, Australia

INTRODUCTION

The surface which occurs when the periodicity of a crystal is terminated has long been recognized as having properties which differ significantly from either the bulk of isolated atomic states. The chemical interest in the area of research dates from the early part of this century and has been concerned mainly with phenomenological, kinetic and thermodynamic investigations of the adsorption of gases onto surfaces, with the obvious implication to the important processes of corrosion and catalysis. During this time, speculation on the microscopic structure of surfaces, at both a geometric and electronic level, was based mainly on chemical intuition derived from experience with the more clearly defined solid and molecular states. In more recent years, and particularly during the last decade, a greater understanding of the geometric and electronic structure of surfaces has been achieved by a variety of techniques which have been designed specifically for the purpose. Of great interest are the experiments which have been based on the scattering of low energy electrons, i.e. with energies of the order 10-1000 eV. These techniques have developed because of the strong interaction of these low energy electrons with matter. For example, electron impact cross sections are typically $\sim 10^{-16}$ cm^2 which is comparable to the dimensions of an atom. This strong interaction with matter persists in the solid state where it can be shown that the penetration depth of low energy electrons into the solid depends on both their kinetic energy and the nature of the target material. Typical values of the penetration depth lie in the range 4-10Å.[1] As a result, the information which is carried by the electrons which have been scattered from solids, will have a significant component which may be attributed to the surface region.

A preliminary division of the types of scattering processes which may occur can be made on the basis of the energy transfer during the interaction. Thus, low energy electron diffraction (LEED) utilizes those electrons which have been elastically back scattered ($\Delta E = 0$ if the trivial recoil of the lattice is neglected) by acquiring momentum from the lattice.[2] In an ordered surface net the coherent back scattering of the elastic electrons provides useful information about the geometric structure of the surface. The alternative process of inelastic scattering, in which energy (and often momentum) is exchanged with the solid, has proved to be a particularly useful technique for determining the electronic structure of the surface region.[3] If the interaction occurs with the atomic core levels, little chemical information is available. Thus, the techniques of Auger electron and ionization spectroscopy are more useful for the elemental analysis of surfaces. If, however, the interaction involves the valence band of the solid, a possibility exists to determine the electronic structure of the surface, with its important implication to both the chemical and physical properties of surfaces. Inelastic electron scattering processes in this category include core VV Auger, loss and appearance potential spectroscopies. A feature which is common to much of this work is that the momentum information which is contained in the scattering process is forfeited by the use of either spatially insensitive electron spectrometers, or polycrystalline surfaces. Indeed, the main research effort has been towards obtaining information about the distribution of energy states in the valence band. In some isolated studies, the momentum of the scattered electron has been measured, in order to characterize the unfilled Bloch states which are responsible for transporting the electron from the solid to the vacuum continuum.[4] Information on the initial momentum distribution of surfaces is non-existent.

In the gas phase, momentum distributions have been obtained by the (e,2e) reaction in which the total kinematics of the inelastic scattering process is measured.[5] With the scattering geometry shown in Fig. 1, an incident electron with kinetic energy E_0 and wavevector $\underline{k}_0$, on interaction with a gaseous atom (or molecule) is scattered so that the two outgoing electrons have kinetic energies E_A and E_B, and wavevectors $\underline{k}_A$ and $\underline{k}_B$. By conservation of energy, $E_0 = E_A + E_B + \zeta$ where ζ is the ionization energy of the ejected electron. Similarly, for the symmetric, non-coplannar geometry $\theta_A = \theta_B = 45°$ and $k_A = k_B$, conservation of momentum leads to an initial momentum of the ejected electron, $\underline{q}$, with a magnitude given by

$$q = \left| (2k_A \cos\theta - k_0)^2 + 4 k_A^2 \sin^2\theta \sin^2\frac{\phi}{2} \right|^{\frac{1}{2}}$$

where ϕ is the out of plane angular variable.

By the use of a pair of directionally sensitive electron spectrometers of the cylindrical mirror type and coincidence counting techniques which select only those pairs of electrons which originate from a single collision event, the angular correlation has been shown to yield the momentum space wave function of the ejected electron. In a gas phase experiment of this kind the gas atoms are randomly oriented so that anisotropic angular correlations are not expected. Since this is not the case in crystalline materials, there exists the intriguing possibility of being able to extend the (e,2e) reaction to solids and to determine the anisotropy of the momentum distribution in well defined crystallographic directions. In addition to determining energy distributions, $N(\zeta)$ versus ζ with incident energies which are not easily obtained with resonance photon sources ($\sim$ 40-1000 eV) the (e,2e) reaction also provides momentum distributions which, for crystals, are expected to be of two kinds. Firstly, the direct determination of the dependences of band energy (ζ) on the momentum transfer, $\underline{q}$, the anisotropic $E(\underline{k})$ dispersion of the delocalized valence band may be determined. Secondly, an angular correlation of the more localized atomic like states, which occasionally overlap the delocalized band, should provide information on the spatial orien-

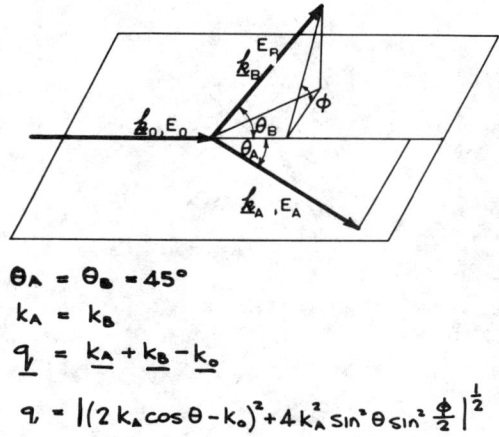

$$\theta_A = \theta_B = 45°$$
$$k_A = k_B$$
$$\underline{q} = \underline{k_A} + \underline{k_B} - \underline{k_o}$$
$$q = \left|(2k_A \cos\theta - k_o)^2 + 4k_A^2 \sin^2\theta \sin^2\frac{\phi}{2}\right|^{\frac{1}{2}}$$

Figure 1. Scattering geometry for the symmetric non coplannar (e,2e) reaction.

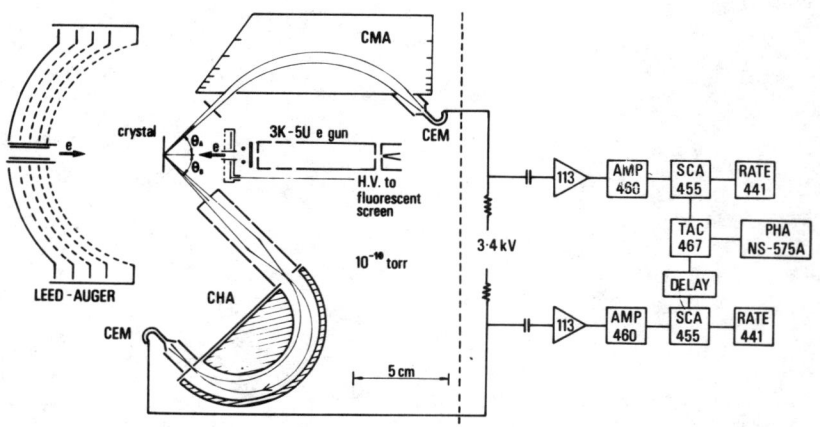

Figure 2. Schematic diagram of the apparatus.

Figure 3. Photograph of the electron gun, CMA and CHA prior to being located in the uhv chamber.

tation of these states. Examples of localized states of this kind are likely to include the dangling bond surface state on semiconductors, impurity levels in the band gap of semiconductors and the relatively tightly bound d orbitals of d^9, d^{10} transition metals.

Furthermore, if the same electron energies which have proved useful in the (e,2e) experiment in gases, (200-800 eV), are applied to the equivalent solid state experiment, the momentum distributions which are obtained should contain a significant contribution from the surface region. The main difficulty in extending the experiment to solids is the fact that the scattering must occur in the forward direction. Thus for thick solids the two outgoing electrons cannot be returned to the vacuum without being further scattered by secondary processes. In the first (e,2e) experiment on solids this problem was overcome by using a thin film target which could be traversed by relatively high energy electrons (~ 15 kV). At these energies the mean free path of the incident electron is of the order 50-100 Å, so that there is always a significant chance of the incident electron undergoing only a single collision event in a film 100-200 Å thick. The difficulty of this approach is that sufficiently thin films, particularly in single crystal form, are usually unavailable. Furthermore, the high incident energies which are required has the effect of reducing the spatial, and therefore the momentum resolution of the experiment, by reducing the range of the angular variable ϕ.

EXPERIMENTAL

During the past year the problem of adapting the (e,2e) reaction to solids has been approached by utilizing the strong back diffracted beams which emerge from crystalline materials. In this work the target used was a tungsten single crystal, which had been spark machined and polished to expose the (001) plane at the surface. Tungsten is particularly suitable for this kind of work since it is highly refractory and therefore easily cleaned by heating to 2200 K. The cleanliness of the surface was routinely monitored by the well established techniques of LEED and Auger electron spectroscopy. The incident electron beam (supplied by a conventional 3K-5U electron gun (Fig. 2) is back diffracted from such a well ordered crystal into highly coherent and monoenergetic beams which only exist in specific directions determined by the crystallography of the solid and the electron energy. For incident electrons with energies greater than 150 eV the penetration into the bulk is sufficient for the diffraction conditions to be determined by the Bragg condition for the bulk, viz. $2\underline{k}_0 \cdot \underline{G} = G^2$ where $\underline{k}_0$ is the incident wavevector in the crystal and $\underline{G}$ is the operating reciprocal lattice vector. With the incident beam at normal incidence $\underline{k}_0 // \underline{G}_{0\ 0\ 2n}$ and reciprocal lattice points of the type 0 0 2n are excited when the wavelength of the electron in the crystal is given by $\lambda = \dfrac{a_0}{n}$ where a_0 is the lattice constant for the tungsten (3.16 Å). The free electron wavelength in vacuum, $\lambda = (\dfrac{150.4}{E(eV)})^{\frac{1}{2}}$, is decreased in the crystal by an amount determined by the inner potential which for tungsten is taken as 16 eV. Thus, for a normal incident electron $\frac{a_0}{n} = (\frac{150.4}{E + 16})^{\frac{1}{2}}$ describes the incident energy (in vacuum) which is necessary to excite a strong specular diffracted beam of the type 0 0 2n which will emerge from the crystal also normal to the surface. Most of the experiments were performed with the 00 10 Bragg beam which was excited with incident energies near 360 eV. By rotating the crystal a few degrees this beam could be displayed on the shielded fluorescent screen which was operated at $\sim$ 1 kV. Direct observation of the diffracted beam allowed optimum tuning of the electron energy to the diffracted condition, in addition to focussing and deflection of the beam to accommodate small misalignments of the electron gun. The operating geometry for the (e,2e) reaction is therefore one in which a strong monoenergetic Bragg beam emerges from the crystal in a well defined direction (viz. normal to the surface) with an intensity of perhaps 10^{-2} of that of the incident beam. It is this beam which becomes the incident beam for a (e,2e) reaction. If such a reaction occurs, the two outgoing electrons which are scattered into the symmetric non-coplannar geometry of the gas phase equipment, are detected by a pair of electron spectrometers as shown in Fig. 2. Coincident events are then detected in the normal way by timing between the pulses from the individual channel electron multipliers (CEM) after they have been routed through separate amplifier and zero cross over discrimination lines. The start pulses are delivered by the cylindrical mirror analyser (CMA) which was set so that $\theta_A = 45°$ and can be rotated out of plane about the axis defined by the incident beam direction such that $-25° < \theta \leqslant 50°$. The resolving power of this spectrometer was measured at $E/\Delta E = 80$. The corresponding stop pulse was delivered by the concentric hemispherical analyser (CHA) which had the feature of an input parallel cylinder electrostatic lens which retarded the electron by a factor of 10 and at the same time imaged the source on the equitorial plane of the hemispheres. This arrangement allowed both high transmission and resolving power, $E/\Delta E = 700$. The CHA could be rotated about the crystal so that $43° < \theta_B < 95°$ and in Fig. 2 is shown rotated 90° about the axis of its parallel cylinder lens for clarity. The gun, in its shield mount, and the two electron spectrometers are shown in Fig. 3 prior to being located in the uhv chamber. Both the CMA and CHA had input aperture stops which allowed a spatial resolution of $\pm 1.5°$. Thus for $E_0 = 360$ eV the resolution of the (e,2e) experiment was $\Delta E \sim 3$ eV and $\Delta k \sim 0.055 Å^{-1}$. The start and stop pulses triggered a time to analog converter and the time spectrum was accumulated in a pulse height analyser. The time resolution of the counting equipment was measured as ~ 3 ns which will be increased for the (e,2e) experiment by a comparable amount due to differing path lengths through either of the spectrometers.

The two outgoing electrons which emerge from the crystal at 45° to the surface normal, will be refracted by the inner potential as they cross into the vacuum continuum. From Snells Law it can be estimated that at 180 eV the electrons are deflected away from the surface normal by a further 2.5°. Therefore, the operating geometry in the crystal will be $\theta_A = \theta_B = 42.5°$ which can be shown to be appropriate for $\zeta = 30$ eV when $\vec{q} = 0$.

For surface studies of this kind it is essential that the surface be maintained in an uhv environment so that contamination from the gas phase can be suppressed. The kinetic theory of gases shows that in a pressure of 10^{-6} torr, for example, a surface will be completely covered in ~ 1 sec if every atom colliding with the surface has a sticking probability of unity. For these experiments, conventional uhv components and procedures were used, and vacua of 10^{-10} torr were routinely obtained. The vacuum chamber with the spectrometers in place was degaussed with a 2000 amp turns coil, and the residual field was nullified to ~ 5 mG with a set of three Helmholz coils.

With the incident beam tuned to the 0010 Bragg beam at 360 eV the operating pass energy of the spectrometers when $0 < \zeta < 20$ eV (170-180 eV) is close to the region of the characteristic triplet of the $N_5 N_{6,7} N_{6,7}$ Auger spectrum of tungsten (160-170 eV). Therefore, as part of the initial evaluation of the apparatus this Auger spectrum was recorded in the usual derivative mode (N(E) versus e) using conventional frequency modulation and phase sensitive detection techniques. With both the CMA and CHA, satisfactory Auger spectra were recorded which indicated that the apparatus was functioning properly. For the (e,2e) experiment, E_0 is necessarily fixed by the Bragg condition so that ζ was selected by simultaneously altering, in tandem, the pass energies of the two spectrometers.

RESULTS

At the time of preparing this report no coincidence events have been detected other than accidental ones, the level of which can be estimated from the adjacent non-coincidence pulse pairs. This result applies to a wide variety of scattering configurations which are listed below:
1. The spectrometer pass energies were fixed to define discrete values of ζ in the range 0-10 and 30-35 eV corresponding to the delocalized valence band and the intense $N_{6,7}$ core levels respectively.
2. As for 1, but continuously scanning the pass energies of the two spectrometers through the required range with a linear voltage ramp generator.
3. 1 and 2 were repeated for a selection of values for the out of plane angular variable ϕ.
4. θ_B was altered in the range 43-60°.
5. The crystal was rotated by up to 5°. 4 and 5 therefore describe an asymmetric coplannar geometry.
6. The experiment was also repeated by tuning the incident beam energy to different specularly diffracted beams at 136 eV (006), 542 eV (0012) and 964 eV (0016).

On one occasion the experiment was run for many hours until the number of counts accumulated in each channel of the PHA (N) was about 10^6. By considering that a true coincidence peak could be observed if its magnitude exceeded the noise level (taken as the standard derivation, $N^{\frac{1}{2}}$) due to accidental coincidences, then less than 1 in 10^3 events are in true coincidence.

DISCUSSION

A major difference between the well established (e,2e) experiment for gases and that described here for the solid state, lies in the respective densities of scatter in the collision zone. In the gas phase, the collision zone will be a cylinder of cross sectional area equal to the scattering cross section, ($\sim 10^{-16}$ cm^2) and a length which is defined by the experimental set-up (~ 3 mm). At a pressure of 10^{-3} torr the probability of an electron colliding with an atom in this zone is only 1 in 10^2. Thus multiple collision events can be safely neglected. However, this is not true of the solid state experiment, where the relevant scattering parameter is the mean free path of the electron, which is typically about 5-10 Å in the energy range used in this study. Thus, the opportunity this provides for multiple processes, combined with the small fraction ($\sim 1\%$) of incident electrons which are specularly diffracted into the required direction for the (e,2e) reaction, must necessarily lead to a higher level of accidental coincidences. Estimating this level is a problem allied to the theory of the secondary electron cascade in metals [6] which has proved to be too difficult to solve quantitatively. However, at a semiquantitative level, it has been considered that below ~ 50 eV scattering occurs via a screened Coulomb potential, and is approximately spherically symmetric (in the centre of mass system) whereas, above this energy, scattering becomes more Rutherford like. In other words, the scatter is strongly peaked in the forward direction which corresponds to a small energy transfer in each collision event. In the gas phase these events are discriminated against by the geometry of the experiment. However, in the solid state there is significant opportunity for the incident electron to undergo many events of this kind, intermixed with elastic scattering processes due to interaction with the lattice, until they are degraded in energy and emerge into the vacuum outside the crystal in the direction selected by the experiment. Clearly, pairs of electrons which are created in this way are not related to a single collision event and therefore cannot be in true coincidence with each other. Nevertheless, on this basis it should be possible, in principle, to see true coincident events from the (e,2e) reaction if the counting statistics are improved until a true coincidence peak can be seen above the background of accidental events. In practice, this has not been achieved with 10^6 counts per channel, and indeed it is virtually impossible to predict how many true coincidence events are likely to be intermixed with the random background due to multiple secondary processes.

REFERENCES

1. P.W. Palmberg, Analytical Chem. 45, 549 (1973).
2. M. Prutton, Metals and Materials 5, 57 (1971).
3. E. Bauer, Vacuum 22, 539 (1972).
4. R.F. Willis, Phys. Rev. Letters 34, 670 (1975).
5. S.T. Hood, I.E. McCarthy, P.J.O. Teubner, and E. Weigold, Phys. Rev. A8, 2494 (1973).
6. P.A. Wolff, Phys. Rev. 95, 56 (1954).

VALIDITY OF THE (e,2e) REACTIONS AS A PROBE OF THE
ATOMIC AND MOLECULAR STRUCTURE

A. Giardini-Guidoni, R. Tiribelli, D. Vinciguerra
Laboratori Nazionali del CNEN, Frascati, Italy

R. Camilloni, G. Stefani
Laboratorio Metodologie Avanzate Inorganiche del CNR,
Via Montorio Romano, Roma, Italy

G. Missoni
Laboratorio Ricerche di Base SNAM Progetti
Monterotondo, Roma, Italy

ABSTRACT

Results of an extensive set of measurement of (e,2e) reactions on noble gases are reported, along with a concise description of the experimental set-up. Some conclusions are drawn on the conditions in which the (e,2e) reaction is a valuable tool to investigate atomic wave functions.

INTRODUCTION

It is by now commonly accepted that, among the various processes introduced by a beam of fast electrons colliding with atomic or molecular targets, those in which ionization occurs and two fast electrons emerge at large angles and are detected in coincidence (the (e,2e) reactions), give valuable information on the electronic properties of atoms and molecules.[1-6] Measurements on (e,2e) reactions have been carried out both on atoms and molecules in a variety of experimental conditions in the few laboratories active in the field[1,2,7,8] however as pointed out in the recent review by McCarthy and Weigold[4] the study of these reactions is still under development, and there is no consensus on the best experimental conditions needed to obtain optimum information on the atomic structure. The possibility of obtaining such information is linked to the validity of the model used in describing the scattering process and in particular to the correct interpretation of the experimental data. In this paper, along with a concise review of the theoretical model and experimental set up, we report an extensive series of measurements carried out on noble gases with the aim of checking the theory and finding experimental conditions in which the (e,2e) reactions can provide reliable information on atomic and molecular structures.

THE PROCESS

The (e,2e) reaction is one in which electron impact ionization takes place and the kinematics of the process is fully determined by measuring the momentum vectors of the initial and final electrons.

In fig. 1 the kinematics of the experiment in the laboratory system is shown: $\bar{p}_0$ is the momentum of the incident electron, $\bar{p}_1$ and $\bar{p}_2$ are the momenta of the two outgoing electrons and $\bar{p}_R$ is the momentum of the residual ion R.

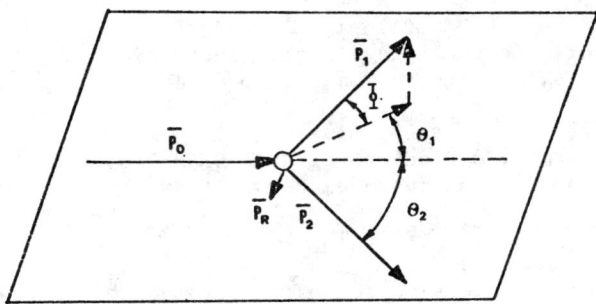

Fig. 1. Schematic representation of an (e,2e) reaction

Conservation of the energy and momenta can be written as:

$$E_o - B_{nl} = E_1 + E_2 + E_R \tag{1a}$$

$$\bar{p}_o = \bar{p}_1 + \bar{p}_2 + \bar{p}_R \tag{1b}$$

where E_o and E_1, E_2 are the kinetic energies of the incident and outgoing electrons and E_R is the recoiling energy. Since the latter is always negligible, the total energy of the final electrons can be related to the binding energy of the initially bound electron B_{nl} by simplifying the relationship (1a):

$$B_{nl} = E_0 - (E_1 + E_2) \tag{2}$$

For such reactions to provide structural information on the atomic target some basic assumption on the way in which the interaction takes place must be made, these assumptions are correlated and are: (1a) the reaction mechanism must be a direct two-body interaction and, secondly, it is assumed that many body effects do not deprive the concept of single particle state of any physical meaning.

If the kinetic energies of the free electrons are made very large compared to the binding energy B_{nl}, and the momentum transfer is also large compared to the characteristic momentum $\langle q \rangle$ of the bound electron, the target binding forces can be expected not to play an important role in the scattering and hence an Impulsive Approximation should apply.

The validity of the second hypothesis depends only on the particular target and can be considered true for closed shell systems such as the noble gases.

To bring the experimental data to a full significance it is further necessary to account for the residual ion interaction which the "interacting electrons" have with the spectator residual ion. It has been shown[9] that, when the atomic potential is a slowly varying function of the coordinates in the main part of the interaction region, the overall matrix element of the process is given in the form:

$$M_{if} = \sum_{|q\rangle} \langle \chi^-_{k_1} \chi^-_{k_2} | t_c | \chi^+_{k_0} q \rangle \langle q \, \Psi^*_R | \Psi_A \rangle \tag{3}$$

in which a two particle transition matrix has been factorized out. Multiple scattering of the incoming and outgoing electrons in the atomic field have been neglected. The $\langle \chi_{k_i} |$ in (3) are electron wave functions which, in a first approximation, are treated as plane waves, while $|t_c|$ is the on shell matrix element for the Coulomb e-e scattering, and Ψ_R and Ψ_A are respectively the residual in ion and atomic wave functions.

According to the I.A. which implies the conservation of the momentum of the residual ion during the interaction, the q value in (3) must be understood as the momentum of the bound electron before interaction. This results in the relation $+\bar{q} = -p_R$. As to the atomic wave functions, if they are expressed by a product of antisimmetrized one electron w.f. (H.F. w.f.) or a C.I. expansion of H.F. w.f.s, the overlap integral $\langle q \, \Psi^*_R | \Psi_A \rangle$ further splits into the Fourier transform $\langle q \, \varphi_{nl}(\bar{r}) \rangle = \varphi_{nl}(\bar{q})$ of the bound electron w.f., which give rise to the observed momentum distribution (o.m.d.) times the overlap integral $\langle \Psi^*_R | \Psi_{A-1} \rangle = (K^I_{nl})^{1/2}$ which accounts for the satellite structure observed in the energy spectra.[2,5]

Calculation of the scattering cross section in this plane wave impulse approximation (P.W.I.A.) leads to the formula:

$$\frac{d^5}{d\Omega_1 \, d\Omega_2 \, dE_1} \alpha \, |\langle k_1 k_2 | t_c | k_0 \, q \rangle|^2 |\varphi_{nl}(\bar{q})|^2 \, K^I_{nl} \tag{4}$$

When the energies of the electrons involved in the scattering are not sufficiently large as compared with the atomic potential, distortions of the electron waves take place. A suitable optical potential[2] is introduced to account for this effect with the result that the $\langle \chi_{k_i} |$ are no longer momentum eigenstates.

These modifications correspond to describing the scattering inside the potential well of the atom, where the electrons momenta are larger than those measured outside. Nevertheless in this

D.W.I.A. the overall matrix element can still be given by the product of a two electron direct scattering matrix, which is now computed off the energy shell, times a distorted momentum distribution. A simple way to compute this distortion has been introduced by McCarthy and coworkers[2,4]; this eikonal approximation gives the following form for the ejected electron w.f.:

$$\chi_k^{\pm} \alpha\, e^{i(1+\beta+i\gamma)\bar{k}\cdot\bar{r}} \tag{5}$$

where $\beta = V/2E$ is a real parameter proportional to the effective potential experienced by the electrons and γ is an attenuation factor mainly affecting the absolute value of the cross section. The effect of β is to change the value of the momentum.

In summary two kinds of spectra can be obtained from the (e,2e) experiment: i) The energy spectrum obtained by varying the incident electron energy while keeping the angles and energies of the outgoing electrons fixed (the peaks observed in this way correspond to the various separation energies of the final ionic states) and ii) The angular distribution obtained at a constant value of the initial and final electron energies by verying the scattering angles. (In this way a particular final ionic state can be selected and, from the shape of the angular distribution the function $|\varphi_{nl}(\bar{q})|^2$ can be deduced).

EXPERIMENTAL

Except for minor modifications the apparatus used in these studies is the same as one previously described.[5] For completeness its features and its performance will be briefly reviewed here. It consists of a stainless steel cylindrical chamber 60 cm high and 130 cm in diameter, connected to a pumping system. All the basic components: the electron gun, the two rotatable electron spectrometers and the gaseous target are placed on the lower flange, which can be easily removed from the chamber. A picture of the flange where it is possible to identify these main components is shown in fig. 2a.

The pump system consists of a rotary mechanical pump, an 8000 l. sec. mercury diffusion pump, and two baffles refrigerated with water and liquid nitrogen respectively. It provides a typical background pressure of 10^{-7} torr. When the gaseous jet is sent into the chamber the pressure rises to about 10^{-6} torr. A He cryopump[11] sitting on the top flange in front of the jet is sometimes used.

The electron gun, a Varian Auger model, provides a well collimated ($\Delta\theta$ about 4.10^{-2} rad.) electron beam in the energy range 50 - 3000 eV. Beam intensity can be varied from a few μA's to 200 μA.

The gaseous beam is obtained by allowing the gas to effuse through a Bendix multichannel array whose thickness is 25 mm. Each channel is 10 μ i.d. and the active area is about 50% of the total. The multichannel array is sealed to the end of an hypodermic needle, .6 mm. i.d., placed about 3 mm below the electron beam path. The intersection of the two beams defines an interaction volume of about $5 \cdot 10^{-3}$ torr.

Figure 2a. Picture of the apparatus for observing (e,2e) reactions. (1) Electron Gun, (2) Faraday cage (3) Electron spectrometers, (4) Multichannel array nozzle.

The two identical electron spectrometers are fixed to rigid supports independently movable on a circular path around the interaction volume on a plane containing the electron beam path. In this coplanar arrangement, the azimuthal angle Φ is equal to zero and the angles θ_1 and θ_2 (see fig. 1) can be independently varied from 28° to 72°. The motion is provided from the outside, by Riber rotary manipulators, allowing for positioning precise to ± 2°. The spectrometers are hemispherical electrostatic analyzers formed by two concentric hemispheres of 105 mm. and 135 mm/diameter respectively. Entrance and exit slits are molybdenum diaphragms with 3 mm i.d. apertures. Fringing fields are reduced by Herzog elements 15 mm. long placed at the entrance and exit of the hemispheres. The energy resolution of each spectrometer can be varied by means of plane decelerating lenses down to a minimum value of 1,5 eV (FWHM). Electron detectors are Bendix channeltron multipliers. The performance of each electron spectrometer was tested by measuring the angular distribution of elastically scattered electrons by some noble gases.[10]

The data acquisition system is very similar to the previously described one[1,10]. Its main components are shown in fig. 2b.

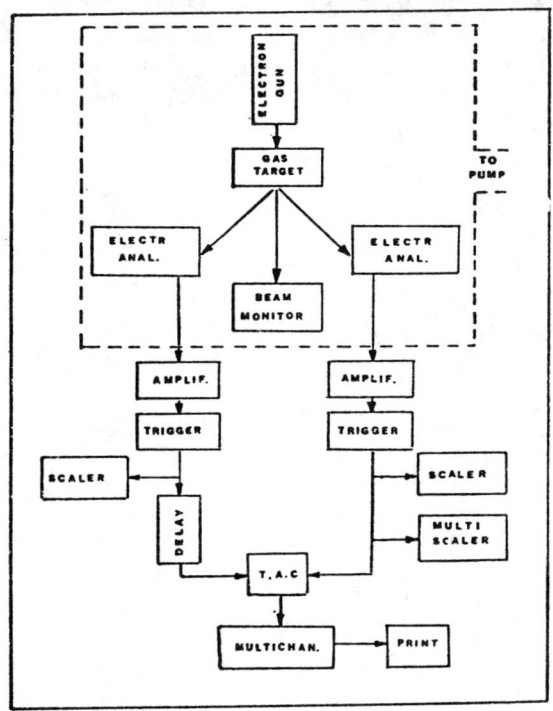

Fig. 2b. Block diagram of the experimental data collection set-up.

RESULTS AND DISCUSSION

In the following we report a series of measurements of angular distributions taken on noble gases for incoming electrons energies ranging from 200 up to 2600 eV. The data are compared with curves computed from the two above mentioned theoretical models P.W.I.A. and D.W.I.A. in the eikonal approximation as given in papers (2-4), where the value of the attenuation factor γ has been set equal to zero, hence neglecting any absorption effect. The β factor = $V/2E$ or, preferably V, has been chosen as a parameter. The H.F. atomic wave functions used in calculating the form factor in the expression (4) have been taken from Clementi and Roetti[12]. The theoretical curves have been computed by folding in the angular and energy resolution of the apparatus with expression (4) and have been normalized not to the area but to the peak value of the experimental distributions, in order to avoid uncertainties connected with the unmeasured tails. In the fig's 4 to 8 we report the measured angular distribution of the electrons ejected in the He 1s ionization, for a number of incident electron energies, together with the theoretical results. All the measurements have been performed in the symmetrical coplanar geometry, by setting $E_1 = (E_o - B_{nl})/2$, $\theta_1 = \theta_2$ so that only values of the momentum q parallel or antiparallel to the direction of the incoming electron are selected through the relationship $\bar{q} = (2p \cos\theta - p_o)\hat{p}_o$ folded in with the apparatus resolution.

The shift and asymmetry of the curves with respect to 45° is due to the strong dependence of the cross section on the scattering angle. According to the peculiar character of the s type orbital, a maximum in the angular distribution, corresponding to a value $q \sim 0$ of the bound electron is clearly shown.

In the low energy range (200 and 400 eV), as can be seen in fig. 4 and 5 both the P.W.I.A. (equivalent to V=0 in D.W.I.A.) and the D.W.I.A. eikonal appear unable to predict the position of the maximum or the width of the angular distribution. The effect of having introduced the parameter V in the D.W.I.A. eikonal is shown in fig. 5, where the curves computed with the values V=20 and 40 eV are reported. The V parameter acts in the sense of increasing the momenta of the interacting electrons and thus affects both the momentum transfer and the momentum balance. However, while the influence on the e-e scattering factor is mainly on the absolute value since its angular shape is only slightly varied by V, the form factor shrinks in the small angle region so that the overall effect is the analogous of having carried out measurements at a higher total energy. By choosing 20 eV as a value of the parameter, the D.W.I.A. eikonal fits the data quite well from 800 eV on, as can be seen in fig. 6. The distortion effects become almost negligible only from 1600 eV on, as shown in fig. 7 and 8. It should be noted that, at least for the He system, the eikonal approximation in which V is equal to 20 eV is almost equivalent to an on shell I.A. description of the process[13].

A summary of the data taken at incident energies 800, and 1600 and 2500 eV is reported in fig. 3. It shows a quite good agreement with the squared Fourier transform of the radial. He 1s electron w.f. as computed by Clementi[12].

At variance with the out of plane geometry for He it is not necessary to use the D.W.I.A. to fit the data even at 200 eV incoming energy. This can be explained by the fact that these experimental conditions are very insensitive to the distortion. This is because the e-e scattering factor is constant and the distortion on the momentum balance introduced by the parameter V is much smaller, since now the change Δq is proportional to $\Delta k_o \sin \Phi/2$ and not to $\Delta k_o \cos \theta$ as in the other geometry.

A behavior analogous but somehow different to the one of He was found for the 2s electrons in Ne. The angular distribution taken at 800 eV disagrees with the one computed in D.W.I.A. eikonal with V=30 as shown in fig. 10. At the higher energies 1600 eV and 2600 eV the eikonal D.W.I.A. appears to be valid. The agreement between the calculated Ne 2s Clementi w.f.[12] and the data is shown in fig. 9 for all the energies investigated.

The angular distribution arising from the Kr 4s electrons has been measured only for an incident electron energy of 800 eV in the coplanar symmetric geometry. This is shown in fig. 13 where it appears that the eikonal approximation with V=20 eV is able to fit the data. In fig. 14 the data have also been compared with the squared Fourier transform calculated by using the Kr 4s Clementi w.f.[12].

For the Xe 5s orbital the best fit to the data (fig. 16 and 17), is obtained in the distorted wave eikonal model, by putting V=10 eV in agreement with the smaller characteristic momentum of the bound electron. In fig. 15 data taken at 400, 800 and 2600 eV incident electron energy are compared with the ρ_{n1}[9] obtained from the Xe 5s Clementi w.f.[12]. From all these data it appears that the characteristic momentum and the binding energy affect the range of validity of the models used in describing the scattering process.

As to the p electrons of Ne, Kr and Xe, the situation is not so simple. The computed shapes never appear to be in such good agreement with the data, even in the high energy range.

In particular in the case of the Ne 2p electrons (figs. 19 to 22), while the intensity ratios and the relative positions of maxima and minima are fairly well reproduced at energies larger than 1500, at 800 eV the agreement fails at angles corresponding to low momentum transfer ($\theta \leq 38°$); both in the coplanar symmetrical ($\theta_1 = \theta_2$) and asymmetrical ($\theta_1 \neq \theta_2$) geometries. All the data are reported in fig. 18 where, as usual, the comparison between measured and calculated squared Fourier transform for the Ne 2p Clementi w.f.[12] is shown.

Data on the ejection of Kr 4p electrons are reported in figs. 23, 24 and 25 along with the comparison with the squared Fourier transform of the Kr 4p Clementi w.f.[12]. From these data it appears that at the lowest energies and small scattering angles ($\theta \leq 42°$) the eikonal approximation cannot explain the data. Better agreement is obtained at higher energies.

Angular correlation of 5p electrons in Xe has been taken only at 400 eV as shown in fig. 26. At this energy the eikonal approximation partially fits the data. At small scattering angles the usual deficit of the measured cross section shows up again. Only the angular correlation obtained is reported and compared with curves belonging to different values of the V parameter for 5p Xe Clementi w.f.[12].

CONCLUSIONS

From the analysis of these results, even if no absolute measurements of the cross section and the spectroscopic factors have yet been carried out, some significant conclusions can be drawn on the check of the reaction mechanism and on the experimental conditions to be chosen to obtain more reliable information on atomic and molecular structure.

In the coplanar geometry the conditions for the validity of the direct interaction without distortion, and for the (e,2e) technique to be a good tool in testing a wave function, are satisfied for incoming electron energies quite high with respect to the binding energy of the orbital under study and for momentum transfer large compared to the $\langle q \rangle$ momentum of the bound electron. In particular the P.W.I.A. model appears suitable to describe the (e,2e) reactions, in case of the external shells, for incoming electron energies of the order of 50 times the binding energy of the orbital and momenta transfer an order of magnitude ($\sim$ 10 times) larger than the $\langle q \rangle$ value.

For incident energies ranging from about 50 to about 20 B_{nl} the eikonal D.W.I.A. can take care of the distortion effects and reliable information on the atomic and molecular orbitals can still be obtained.

For lower incident energies ($E_o \sim 10\ B_{nl}$) a more sophisticated model must be used in describing the scattering, so that, at this stage of the theory, experiments in this range of energy appear to be more useful in obtaining information about the reaction mechanism than in testing atomic w.f.'s.

REFERENCES

1. R. Camilloni, A. Giardini Guidoni, G. Stefani and R. Tiribelli, Phys. Rev. Letters 29, 618 (1972).
2. S. T. Hood, I. E. McCarthy, P. J. O. Tuebner and E. Weigold, Phys. Rev. A8, 2494 (1973).
3. A. Ugbabe, E. Weigold and I. E. McCarthy, Phys. Rev. A11, 576 (1975); E. Weigold, S. T. Hood and I. E. McCarthy, Phys. Rev. A11, 566 (1975).

4. I. E. McCarthy and E. Weigold, to be published in Phys. Rep.
5. A. Giardini Guidoni, G. Missoni, R. Camilloni and G. Stefani, Electron and Photon Interactions with Atoms (Plenum Press, 1976, H. Kleinpoppen and M. R. C. McDowell, editors), p. 194.
6. G. Neudachin, G. A. Novoskol'tseva, Yu. F. Smirnov, Sov. Phys. JETP $\underline{28}$, 540 (1969).
7. S. T. Hood, A. Hammnett and C. E. Brion, Chem. Phys. Lett. $\underline{39}$, 252 (1976).
8. K. Jung, E. Schubert, D. A. L. Paul and H. Ehrhardt, J. Phys. B. Atom. Molec. Phys. $\underline{8}$, 1330 (1975).
9. J. P. Coleman, Case Studies in Atomic Collision Physics I (North Holland, 1969, E. W. McDaniel and M. R. C. McDowell, editors), p. 101.
10. R. Camilloni, G. Stefani, A. Giardini Guidoni, R. Tiribelli and D. Vinciguerra, Proceeding of the 5th Congress on Vacuum Science and Technology, Perugia (1975); A. Giardini Guidoni, R. Camilloni, G. Stefani, R. Tiribelli, D. Vinciguerra and E. Weigold, IX I.C.P.E.A.C., Seattle (1975), Abstract of papers, p. 490, R. Camilloni, A. Giardini Guidoni, G. Stefani, R. Tiribelli and D. Vinciguerra, Chem. Phys. Letters (1976) in press.
11. G. Baldacchini, Cryogenics $\underline{14}$, 574 (1974).
12. E. Clementi and C. Roetti, Atomic Data and Nuclear Data Tables (Academic Press, 1974) Vol. 14.

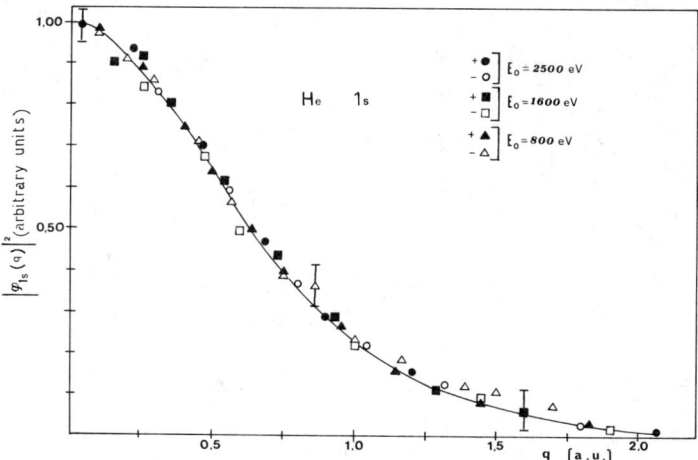

Fig. 3. q distribution resulting in D.W.I.A. (eikonal V = 20 eV) in coplanar symmetrical conditions for incident electron energies of 800, 1600 and 2500 eV. The full curve is the squared Fourier transform of the He 1s Clementi wave function to which the data have been normalized. (Open marks are relative to q antiparallel to the momentum of the incoming electron). B_{nl} optical = 24.58 eV.

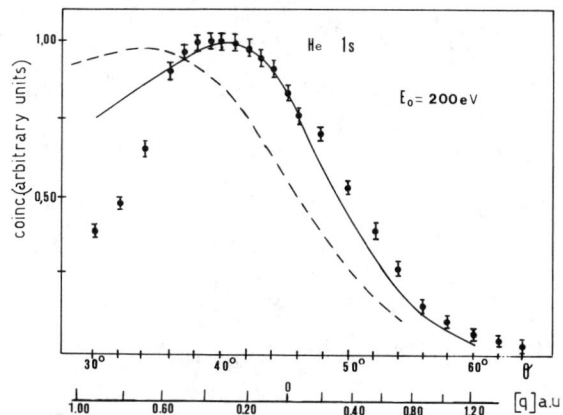

Fig. 4. Angular correlation for the 1s orbital of He. Measured coincidence rate is compared with shapes calculated for the 1s Clementi wave function in P.W.I.A. (-------) and D.W.I.A. (eikonal V = 20 eV)(————). The correspondence between scattering angle and [q] value determined from V = 20 eV is also reported.

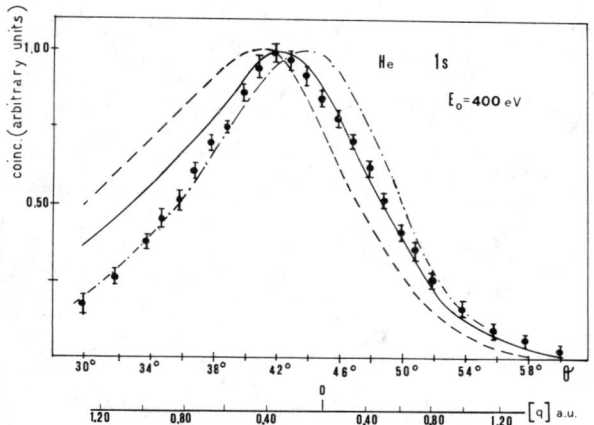

Fig. 5. Angular correlation for the 1s orbital of He. Measured coincidence with shapes calculated for the 1s Clementi wave function in P.W.I.A. (-------) and D.W.I.A. (eikonal V = 20 eV)(————) and V = 40 eV (-··-··-··-). The correspondence between scattering angle and [q] value determined from V = 20 eV is also reported.

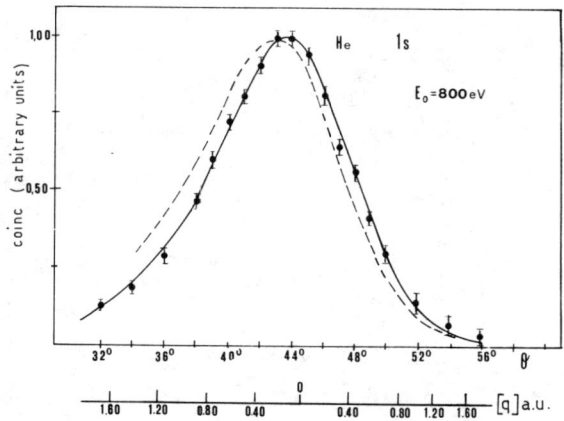

Fig. 6. Angular correlation for the 1s orbital of He. Measured coincidence rate is compared with shapes calculated for the 1s Clementi wave function in P.W.I.A. (-------) and D.W.I.A. (eikonal V = 20 eV)(————). The correspondence between scattering angle and [q] value determined from V = 20 eV is also reported.

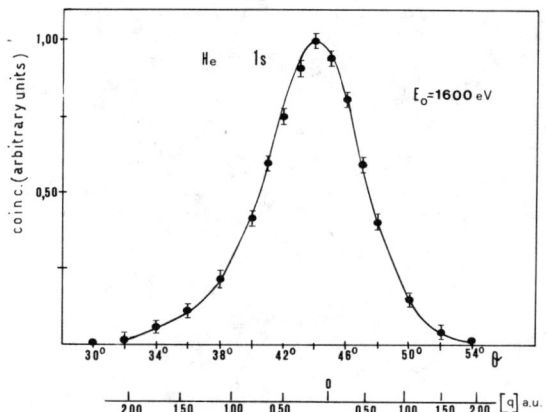

Fig. 7. Angular correlation for the 1s orbital He. Measured coincidence rate is compared with shapes calculated for the 1s Clementi wave function in D.W.I.A. (eikonal V = 20 eV) (———) indistinguishable from P.W.I.A. The correspondence between scattering angle [q] value determined from V = 20 eV is also reported.

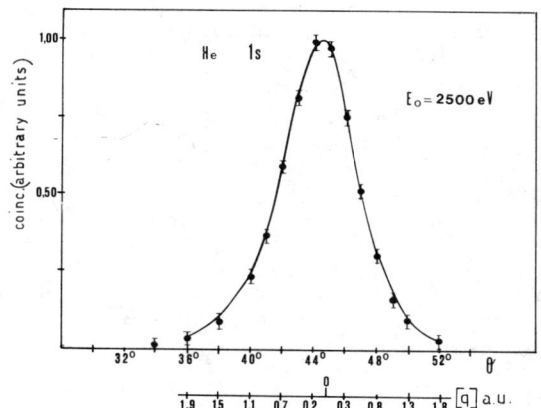

Fig. 8. Angular correlation for the 1s orbital of He. Measured coincidence rate is compared with shapes calculated for the 1s Clementi wave function D.W.I.A. (eikonal V = 20 eV) (———) which is indistinguishable from the P.W.I.A. result). The correspondence between scattering angle and [q] value determined from V = 20 eV is also reported.

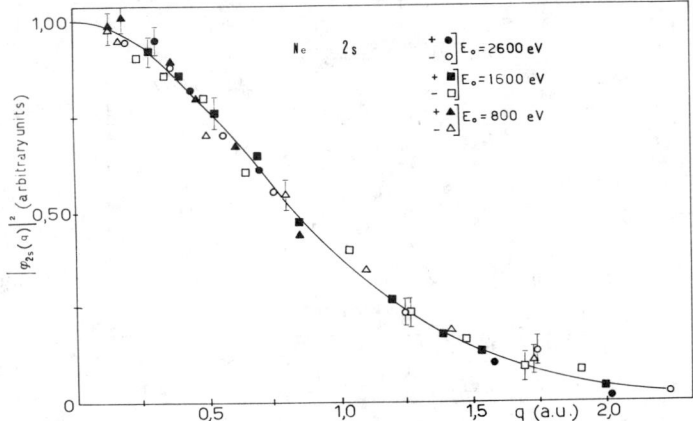

Fig. 9. q distribution resulting from D.W.I.A. calculation (eikonal V = 30 eV) for coplanar symmetric conditions for incident electron energies of 2600, 1600 and 800 eV (for $\theta > 42°$). The full curve is the squared Fourier transform of the Ne 2s Clementi w.f. to which the data have been normalized. (Open marks are relative to $[q]$ antiparallel to the momentum of the incoming electron). B_{nl} optical = 48.5 eV.

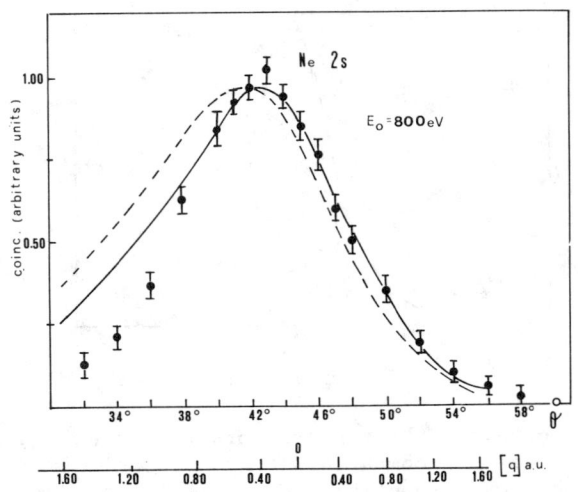

Fig. 10. Angular correlation for the 2s orbital of Ne. Measured coincidence with curves calculated for the Ne 2s Clementi wave function in the P.W.I.A. (-------) and D.W.I.A. (eikonal V = 30 eV)(————) approximations. The correspondence between scattering angle and $[q]$ value as determined from V = 30 eV is also reported in the lower scale.

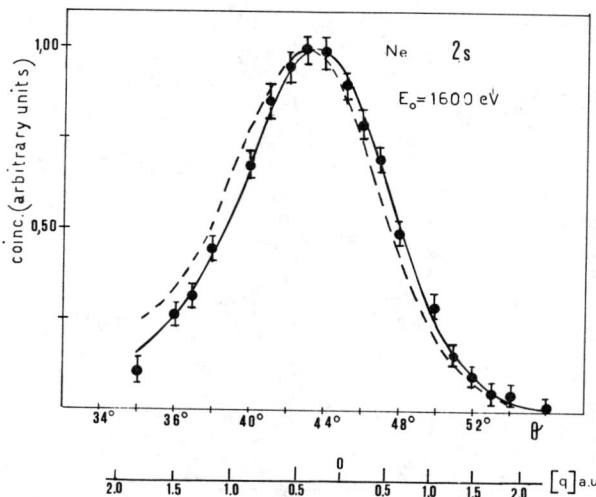

Fig. 11. Angular correlation for the 2s orbital of Ne. Measured coincidence rate is compared with curves calculated for the Ne 2s Clementi wave function in P.W.I.A. (-------) and D.W.I.A. approximations (eikonal V = 30 eV (———)). The correspondence between scattering angle and [q] value determined from V = 30 eV is also reported in the lower scale.

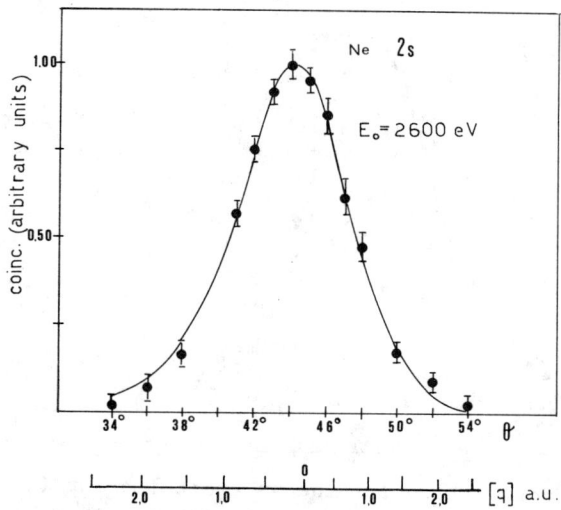

Fig. 12. Angular correlation for the 2s orbital of Ne. Measured coincidence rate is compared with curves calculated for the Ne 2s Clementi wave function in D.W.I.A. (eikonal V = 30 eV almost indistinguishable from P.W.I.A.). The correspondence between scattering angle and [q] value determined from F = 30 eV is also reported in the lower scale.

220

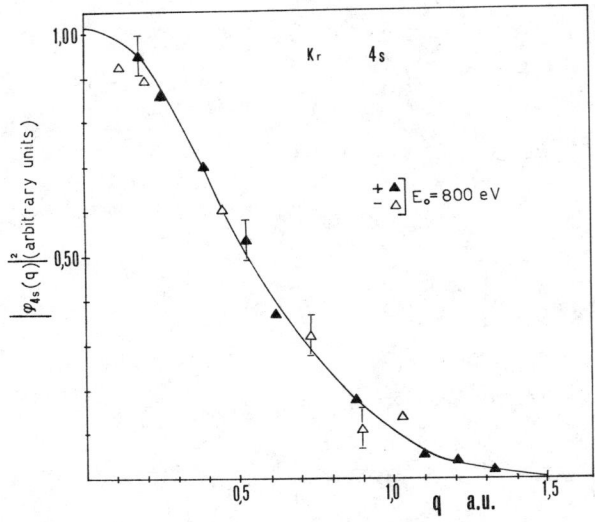

Fig. 13. q distribution resulting from D.W.I.A. calculation (eikonal V = 20 eV) for incident electron energy of 800 eV. The full curve is the squared Fourier transform of the Kr 4s Clementi w.f. to which the data have normalized. B_{nl} optical = 27.51 eV.

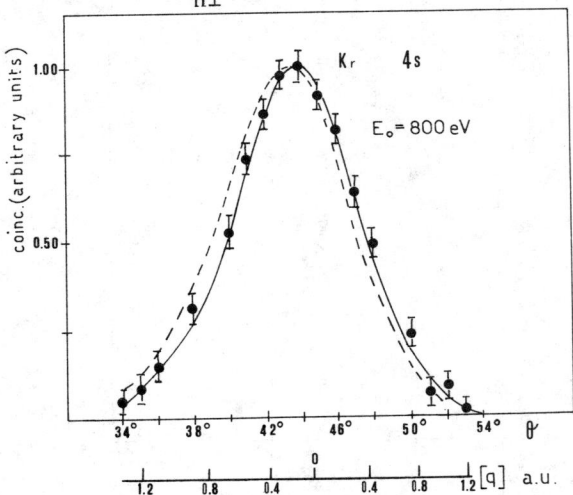

Fig. 14. Angular correlation for the 4s orbital of Kr (B_{nl} = 27.5 eV). Measured coincidence rate is compared with curves calculated for the Kr 4s Clementi wave function in P.W.I.A. (-------) and D.W.I.A. (————) (eikonal V = 20 eV approximations). The correspondence between scattering angle and [q] values determined from V = 20 is also reported.

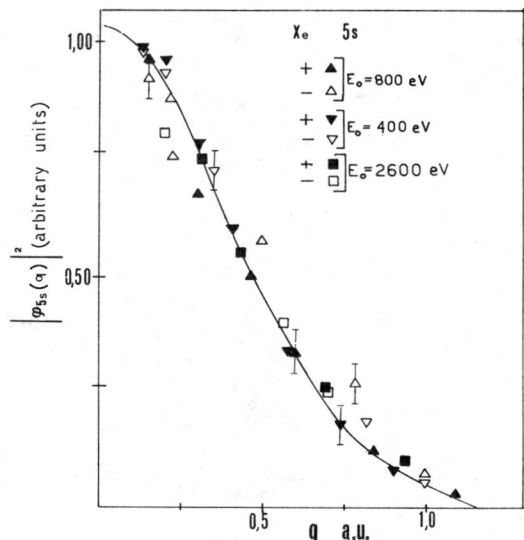

Fig. 15. q distribution resulting from D.W.I.A. calculation (eikonal V = 10 eV) for coplanar symmetric conditions for incident electron energies of 400, 800 and 2600 eV. The full curve is the squared Fourier transform of Xe 5s double zeta Clementi w.f. to which the data have been normalized. (Open marks are relative to q antiparallel to the momentum of the incoming electron). B_{nl} optical = 23.40 eV.

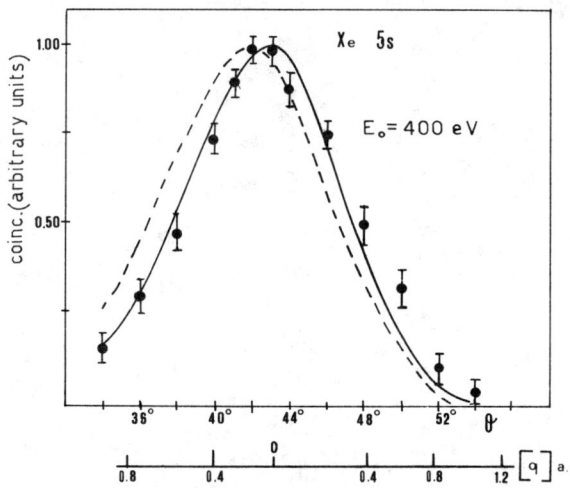

Fig. 16. Angular correlation for the 5s orbital of Xe. Measured coincidence rate is compared with curves calculated for the Xe 5s Clementi wave function in P.W.I.A. (-------) and D.W.I.A. approximations (————) (eikonal V = 10 eV). The correspondence between scattering angle and [q] value is in the lower scale.

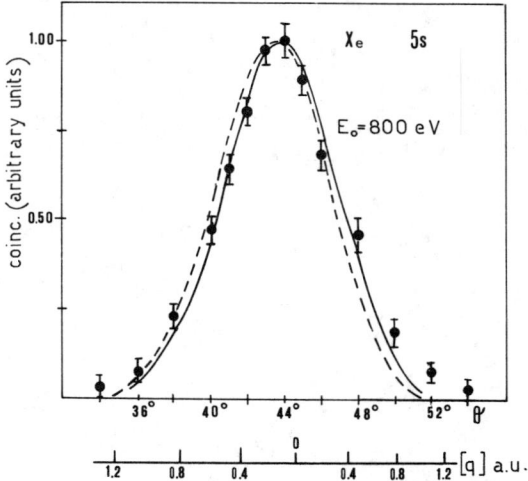

Fig. 17. Angular correlation for the 5s orbital of Xe. Measured coincidence rate is compared with curves calculated using the Xe 5s Clementi wave function in the P.W.I.A. (-------) and D.W.I.A. approximations (———) (eikonal V = 10 eV). The correspondence between scattering angle and [q] value is shown in the lower scale.

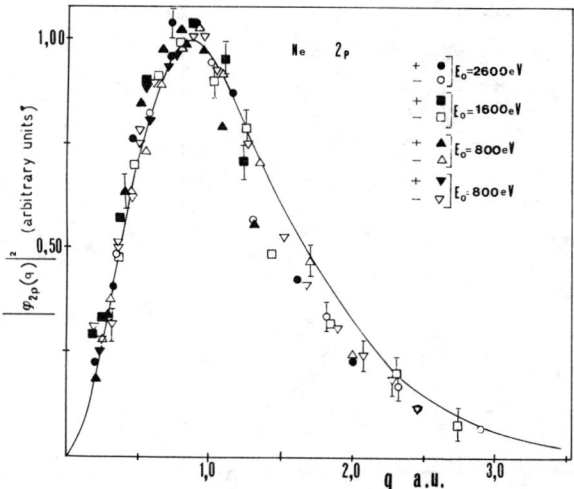

Fig. 18. q distribution from D.W.I.A. calculation (eikonal V = 10 eV) for the coplanar symmetric and asymmetric conditions for incident electron energies of 800, 1600 and 2600 eV. The full curve is the squared Fourier transform of the Ne 2p Clementi w.f. to which the data have been normalized. (Open marks are relative to [q] antiparallel to the momentum of the incoming electron). B_{nl} optical = 21.59 eV.

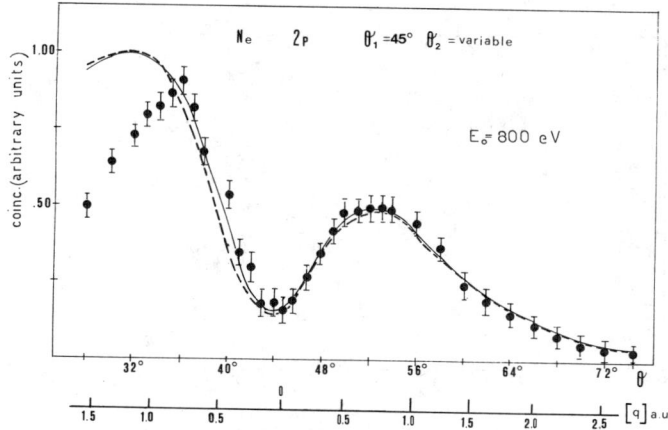

Fig. 19. Angular correlation for the 2p orbital of Ne measured in coplanar asymmetric geometry for the incident electron energy of 800 eV. Data are compared with curves calculated in P.W.I.A. (-------) and D.W.I.A. approximations (————) (eikonal V = 10 eV) for the 2p Ne Clementi wave function.

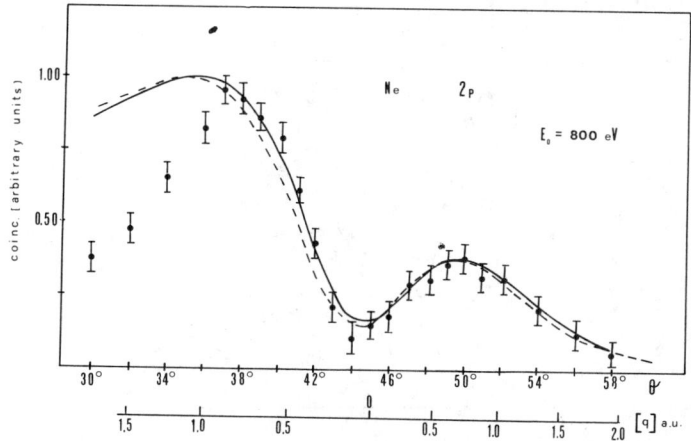

Fig. 20. Angular correlation for the 2p orbital of Ne measured in coplanar symmetric geometry at an incident electron energy of 800 eV. Data are compared with shapes calculated using the P.W.I.A. (-------) and D.W.I.A. theories (————) (eikonal V = 10 eV) for the 2p Ne Clementi wave function.

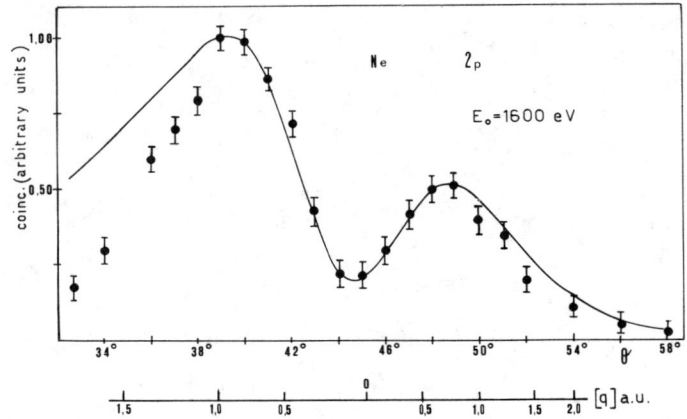

Fig. 21. Angular correlation for the 2p orbital of Ne measured in coplanar symmetric geometry at an incident electron energy of 1600 eV. Data are compared with a curve calculated for the 2p Ne Clementi w.f. in the P.W.I.A. approximation which is almost indistinguishable from D.W.I.A. results (eikonal V = 10 eV).

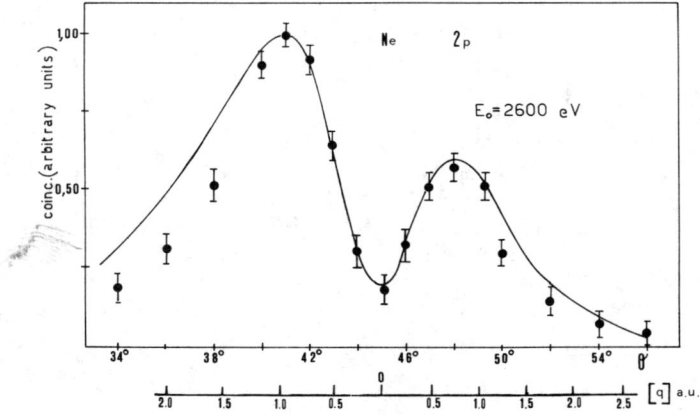

Fig. 22. Angular correlation for the 2p orbital of Ne measured for coplanar symmetric conditions at an incident electron energy of 2600 eV. Data are compared with a curve calculated for the 2p Ne Clementi w.f. in the P.W.I.A. approximation which is almost indistinguishable from D.W.I.A. results (eikonal V = 10 eV).

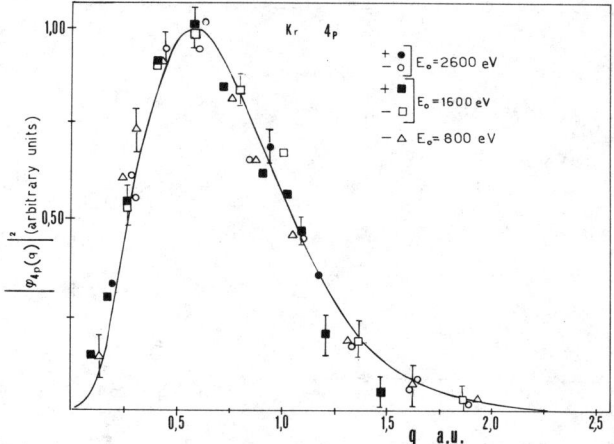

Fig. 23. q distribution for the 4p electron in Kr resulting from D.W.I.A. calculations (eikonal V = 15 eV) for the coplanar symmetric conditions for an incident electron energy of 800 ($\theta > 42°$), 1600 and 2600 eV. B_{nl} optical = 14.00 eV.

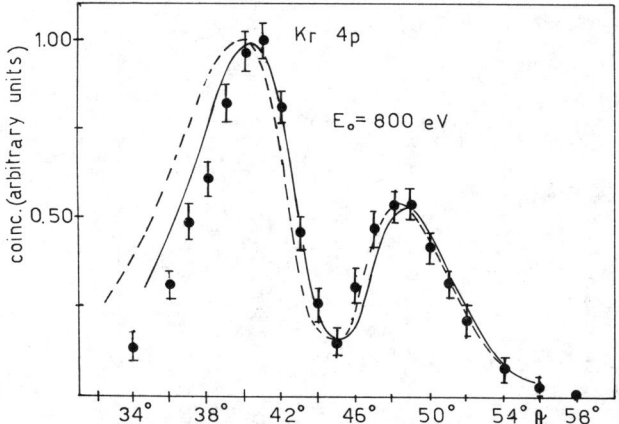

Fig. 24. Angular correlation for the ejection of 4p electrons in Kr at an incident electron energy of 800 eV. Comparison is made with curves calculated from the P.W.I.A. (-------) and eikonal approximations with V = 15 eV (————) for the Kr 4p Clementi w.f.

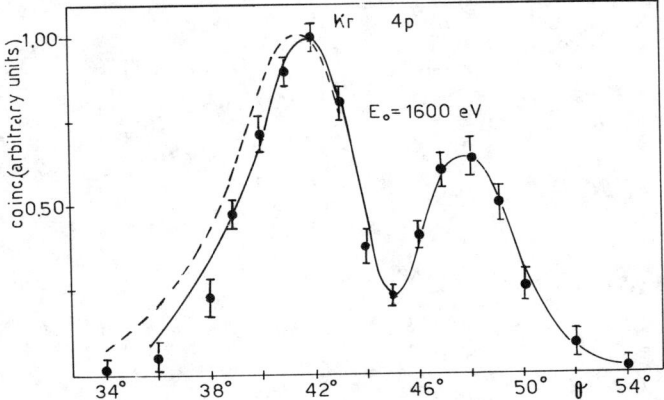

Fig. 25. Angular correlation for the 4p orbital in Kr for an incoming electron energy of 1600 eV compared with a calculated curve obtained by use of P.W.I.A. (-------) and D.W.I.A. approximations (————) (eikonal V = 15 eV) for the Kr 4p Clementi w.f.

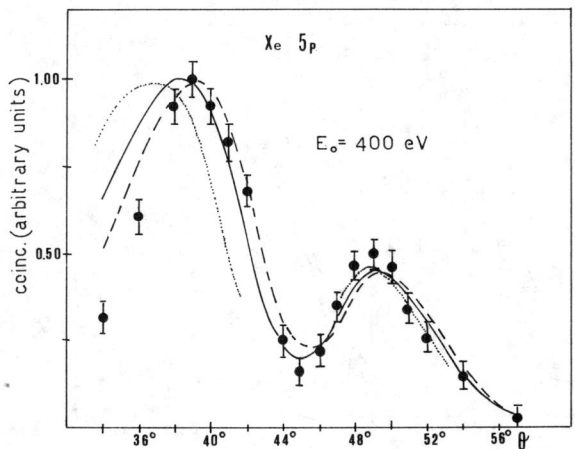

Fig. 26. Angular correlation observed in the ejection of Xe 5p electrons at an incident energy of 400 eV. Data are compared with curves calculated in the P.W.I.A. (.......) and eikonal approximations for V = 10 eV (————) and V = 20 eV (-------) for the Xe 5p double zeta Clementi w.f. B_{nl} optical = 12.13 eV.

DISTORTED WAVE CALCULATIONS FOR (p,2p) REACTIONS*

R. D. Koshel
Ohio University, Athens, Ohio 45701

In this presentation I wish to discuss the theoretical work that my collaborators and I are presently doing and are planning to do in a study of (p,2p) reactions using the distorted wave impulse approximation (DWIA). Since some of this work is only in its preliminary stages there will not be large numbers of slides showing a comparison between experiment and calculation. I will show the results of a few calculations; however, it will be my primary purpose to discuss various aspects of our work which involve what I believe to be the major uncertainties in such a study. As a result I will discuss in varying degrees and not in the order shown the topics of:
1. Distorted wave effects in (p,2p) reaction.
2. Off-the-energy shell behaviour of the nucleon-nucleon t matrix.
3. The calculation of the optical model from the nucleon-nucleon interaction.
4. Multistep processes.

The work I will discuss here is based on the knockout formalism developed by Kazaks and Koshel.[1] This study retains only the lowest order term of this formalism such that the transition matrix element for the (p,2p) reaction can be written as

$$T_{fi} = (N/2)^{1/2} \langle \eta(p_1,p_2)\Phi_R | T | \chi_o(p_1)\Phi_1(p_1)\Phi_T \rangle \\ - (N-1)(N/2)^{1/2} \langle \eta(p_r,p_2)\Phi_R^r | T_{ex} | \chi_o(p_1)\Phi_1(p_1)\Phi_T \rangle, \qquad (1)$$

where the quantities η are the antisymmetrized functions

$$\eta(p_i,p_j) = \chi_1(p_i)\chi_2(p_j)\Phi_1(p_i)\Phi_2(p_j) - \chi_1(p_j)\chi_2(p_i)\Phi_1(p_j)\Phi_2(p_i). \qquad (2)$$

The χ functions are the plane-wave states of the protons and the Φ functions are the internal wave functions of the appropriate particle. The subscripts T and R appearing on the Φ functions denote, respectively, the target and residual nucleus. For the protons these are the spin and isospin functions. N is the number of relevant protons in the target. The terms have been correctly symmetrized with respect to the exchange of the two outgoing protons and the exchange of the incident particle with one of the target protons. This target exchanged proton is symbolized by the letter r. Figure 1 shows the relevant coordinates and interactions. For the lowest order one step process the operator T is given by

$$T = \Omega_{12}^{(-)} t(E) \Omega_1^{(+)} + \Omega_{12}^{(-)} V_1. \qquad (3)$$

$\Omega_1^{(+)}$ and $\Omega_{12}^{(-)}$ are, respectively, the distortion operators for the

*Work supported in part by the National Science Foundation.

initial and final states. The operator t(E) is given by the expression

$$t(E) = v_{12} + t(E)G_o(E)v_{12}, \quad (4)$$

where v_{12} is the proton-proton interaction and $G_o(E)$ is the free-particle Green operator for the two outgoing protons with the residual nucleus acting as a spectator, i.e.,

$$G_o(E) = (E^{(+)} - K_1 - K_2 - H_R)^{-1}. \quad (5)$$

Here $E^{(+)}$ denotes $E+i\varepsilon$, where the limit $\varepsilon \to o^+$ is meant to be taken. H_R is the internal Hamiltonian of the residual nucleus. K_1 and K_2 are the kinetic energy operators of the two protons in the center of mass system, i.e., K_1 describes the motion of the incident proton with respect to the target and K_2 describes the relative motion of the bound proton to the residual nucleus.

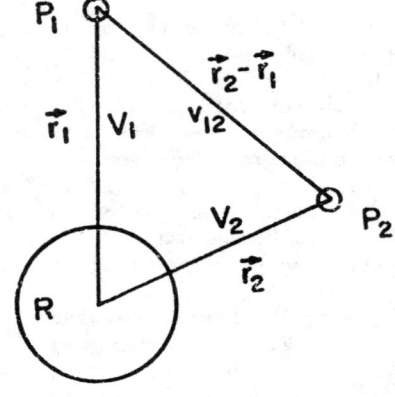

Fig. 1. The relevant coordinates and interactions.

The second term in Eq.(2) is a recoil term which describes the interaction of the incident proton with the residual nucleus with the emitted proton acting as a spectator.

The exchange transition operator T_{ex} is given by

$$T_{ex} = \Omega_{r2}^{(-)} t_{ex}(E) \Omega_1^{(+)} + \Omega_{r2}^{(-)} (V_1 + v_{12} - v_{r2}), \quad (6)$$

where $\Omega_{r2}^{(-)}$ is also a distortion operator for the exit channel except that it now describes the motion of the exchanged proton. The operator $t_{ex}(E)$ is given by

$$t_{ex}(E) = v_{12} + t_{r2}(E)G_o(E)v_{12}, \quad (7)$$

where

$$t_{r2}(E) = v_{r2} + t_{r2}(E)G_o(E)v_{r2}. \quad (8)$$

It is obvious why this description is denoted the distorted wave impulse approximation; we have our initial and final states altered by the distortion operators and the transition operators do not have an interaction between the emitted particle and the residual nucleus. These results obtained here are almost identical to those used by McCarthy in his study of (e,2e) reactions.[2]

I would now like to say something about multistep processes. Let us look at the expression for the distortion operator for the incident channel. It is given by

$$\Omega_1^{(+)} = (1 + G_M V_1), \qquad (9)$$

where

$$G_M = (E^{(+)} - K_1 - K_2 - V_1 - V_2 - H_R)^{-1}. \qquad (10)$$

Thus we have

$$\Omega_1^{(+)} |\chi_o(p_1) \Phi_1(p_1) \Phi_T\rangle = |\psi_o^{(+)}(p_1) \Phi_1(p_1) \Phi_T\rangle, \qquad (11)$$

where $\psi_o^{(+)}$ is the distorted state and is usually taken to be the state which describes elastic scattering from the target nucleus. This is true if V_1 is taken to be a potential which does not allow excitations of the target nucleus. This does not have to be the case. The result of the operation of $\Omega_1^{(+)}$ on the plane-wave state could allow for the excitation of the target nucleus, in particular, the excitation of rotational and vibrational states. Thus our distorted states would be coupled channel states. For the transition operators used in this work we call such a process a coupled channel impulse approximation (CCIA). Recently we have finished a coupled channel Born approximation (CCBA) code[3] which allow for finite range and recoil effects in the study of nuclear transfer reactions. We will extend this work to the CCIA and hopefully present the results of such calculations in the near future.

For this presentation we will neglect the second term in the expression given in Eq.(1). We call this term the knockout exchange term. At present we have not attempted to calculate this term but we hope to do so in the near future. We also neglect the recoil term, that is the second term in Eq.(2). We have calculated this term in one case and have found it to be small[4]. For the sake of clarity I will not include the exchange of the two outgoing protons. This effect is included in our calculations but it is not necessary for our discussion.

Thus what we will investigate is

$$\begin{aligned} T_{fi} &= N^{1/2} \langle \chi_1(p_1) \chi_2(p_2) \Phi_1(p_1) \Phi_2(p_2) \Phi_R | T | \chi_1(p_1) \Phi_1(p_1) \Phi_T \rangle \\ &= N^{1/2} \langle \psi_{12}^{(-)}(p_1,p_2) \Phi_1(p_1) \Phi_2(p_2) \Phi_R | t(E) | \psi^{(+)}(p_1) \Phi_1(p_1) \Phi_T \rangle, \end{aligned} \qquad (12)$$

where the functions $\psi^{(+)}$ and $\psi^{(-)}$ are the distorted waves with the appropriate boundary conditions.

In general one cannot decouple $\psi_{12}^{(-)}(p_1,p_2)$ into a product of two distorted waves. If one, however, assumes the infinite mass approximation for the residual nucleus this can be accomplished. We make this (static) approximation for the present discussion. It is possible to do better than the static approximation by means of semi-classical arguments. This feature is included in our knock-out code.

The method we have used to perform the calculations is the plane-wave expansion method developed by Robson and Koshel.[5] This method has proven to be extremely useful when used in the analysis of other nuclear reactions. Let me now briefly review this method.

If we have an optical model wave function given by

$$\psi(\vec{k},\vec{r}) = 4\pi \sum_{LM} i^L \frac{u_L(kr)}{kr} Y_{LM}^*(\hat{k}) Y_{LM}(\hat{r}), \quad (13)$$

we can write this as

$$\psi(\vec{k},\vec{r}) = \sum_{nLM} a_{nL} Y_{LM}^*(\hat{k}) \int d\hat{k}_n \exp(i\vec{k}_n \cdot \vec{r}) Y_{LM}(\hat{k}_n), \quad (14)$$

provided we use the expansion

$$\frac{u_L(kr)}{kr} = \sum_n a_{nL} j_L(k_n r). \quad (15)$$

We have shown that this generalized Schlömilch expansion is uniformly convergent.[6]

If we now make use of this plane-wave expansion method our expression for the transition matrix element becomes

$$T_{fi} = N^{\frac{1}{2}} \sum a_{n_1 L_1} a_{n_2 L_2} a_{n_o L_o} Y_{L_1 M_1}(\hat{k}_1) Y_{L_2 M_2}(\hat{k}_2) Y_{L_o M_o}^*(\hat{k}_o)$$

$$\times \iiint d\hat{k}_{n_1} d\hat{k}_{n_2} d\hat{k}_{n_o} Y_{L_1 M_1}(\hat{k}_{n_1}) Y_{L_2 M_2}^*(\hat{k}_{n_2}) Y_{L_o M_o}(\hat{k}_{n_o}) \quad (16)$$

$$\times \langle \vec{k}_{n_1} \vec{k}_{n_2} \Phi_1(P_1) \Phi_2(P_2) \Phi_R | t(E) | \vec{k}_{n_o} \Phi_1(P_1) \Phi_T \rangle,$$

where the summation is over n_o, n_1, n_2, L_o, L_1, L_2, M_o, M_1, and M_2. We have neglected all spin-orbit effects in the scattering states. The matrix element in this expression which describes the transition is a plane-wave matrix element. We now concentrate our attention on this matrix element. We define

$$ME = N^{\frac{1}{2}} \langle \vec{k}_{n_1} \vec{k}_{n_2} \Phi_1(P_1) \Phi_2(P_2) \Phi_R | t(E) | \vec{k}_{n_o} \Phi_1(P_1) \Phi_T \rangle \quad (17)$$

$$= N^{\frac{1}{2}} \langle \vec{k}_{n_1} \vec{k}_{n_2} \Phi_1(P_1) \Phi_2(P_2) | t(E-E_R) | \vec{k}_{n_o} \Phi_1(P_1) (\Phi_R, \Phi_T) \rangle,$$

where (Φ_R, Φ_T) is the incomplete scalar product of the two state vectors and

$$t(E-E_R) = v_{12} + t(E-E_R) G_o(E-E_R) v_{12}, \quad (18)$$

with

$$G_o(E-E_R) = (E^{(+)} - E_R - K_1 - K_2)^{-1}. \quad (19)$$

E_R is the energy of the residual nucleus.

In order to bring in more clearly the nuclear structure information we expand Φ_T as

$$\Phi_T = \Sigma (JjMm|J_T M_T)(Tt_2 T_z t_{z2}|T_T T_{zT}) A(iJT;n\ell j;J_T T_T) \\ \times \Phi^i_{JM,TT_z} \Sigma (\ell s m_\ell m_s|jm) u^i_{n\ell jm_\ell}(\vec{r}_2) \Phi_{sm_s,t_2 t_{z2}}(p_2), \quad (20)$$

where Φ^i_{JM,TT_z} is an eigenfunction of H_R with angular momentum J, projection M, isospin T and isospin projection T_z. The superscript i is another quantum number used to differentiate between states with the same quantum numbers J, M, T, and T_z. The function $u^i_{n\ell jm_\ell}$ is the wave function of the bound proton. The quantities $A(iJT;n\ell j;J_T T_T)$ are the expansion coefficients. The first summation is over the quantities i, n, ℓ, j, m, J, T, and M. The second is over m_ℓ and m_s. If we insert this expression into the one for ME we arrive at

$$ME = (T_R t_2 T_{zR} t_{z2}|T_T T_{zT}) \Sigma (J_R j M_R m|J_T M_T)(\ell s m_\ell m_s|jm) \quad (21) \\ \times S^{\frac{1}{2}}(n\ell j) <\vec{k}_{n_1}\vec{k}_{n_2}(12)|t(E-E_R)|\vec{k}_{n_o} u_{n\ell jm_\ell}(12)>,$$

where

$$S^{\frac{1}{2}}(n\ell j) = N^{\frac{1}{2}} A(J_R T_R;n\ell j;J_T T_T) \quad (22)$$

is the spectroscopic amplitude. We have omitted the symbol i because we assume we know which state we have for the residual nucleus. The notation (12) appearing in Eq.(20) indicates the spin and isospin functions of the two protons.

We now denote the matrix element appearing in Eq.(21) as TME. If we take the Fourier transform of the bound state wave dunction we have

$$TME = (2\pi)^{-3/2} \int d\vec{k}\, g_{n\ell jm_\ell}(\vec{k}) \int d\vec{r}_1 d\vec{r}_2 d\vec{r}_3 d\vec{r}_4 \exp(-i\vec{k}_{n_1}\cdot\vec{r}_1)\exp(-i\vec{k}_{n_2}\cdot\vec{r}_2) \\ \times <\vec{r}_1\vec{r}_2(12)|t(E-E_R)|\vec{r}_3\vec{r}_4(12)>\exp(i\vec{k}_{n_o}\cdot\vec{r}_3)\exp(i\vec{k}\cdot\vec{r}_4), \quad (23)$$

where we have gone over to the coordinate representation. It should be mentioned here that we do not make a local approximation for the t-matrix. We now take

$$K_1 + K_2 = K_{cm} + K_{rel} \quad (24)$$

where K_{cm} is the kinetic energy operator for the center of mass motion of the two protons and K_{rel} is the kinetic energy operator for the relative motion. This is inserted in the equation for $t(E-E_R)$. If we then transform Eq.(23) to relative and center of mass coordinates it is quite easy to show that

$$TME = (2\pi)^{3/2} g_{n\ell j m_\ell}(-\vec{q})$$

$$\times \langle \vec{k}_{n_1} - (m/M)\vec{Q}_{12}(12) | \hat{t}(E-E_R-\hbar^2 Q_{12}^2/2M) | \vec{k}_{n_0} - (m/M)\vec{Q}_{12}(12) \rangle \quad (25)$$

where

$$\vec{Q}_{12} = \vec{k}_{n_1} + \vec{k}_{n_2}, \quad (26)$$

$$\vec{q} = \vec{k}_{n_0} - \vec{k}_{n_1} - \vec{k}_{n_2}, \quad (27)$$

and $M = 2m$. We also have

$$\hat{t}(E-E_R-\hbar^2 Q_{12}^2/2M) = v_{12} + \hat{t}(E-E_R-\hbar^2 Q_{12}^2/2M)$$

$$\times (E^{(+)}-E_R-\hbar^2 Q_{12}^2/2M - K_{rel})^{-1} v_{12}. \quad (28)$$

We can also write Eq.(25) as

$$TME = \langle \vec{k}_{n_1} + \vec{k}_{n_2} | u_{nj\ell m_\ell} | \vec{k}_{n_0} \rangle$$

$$\times \langle \vec{k}_{n_1} - (m/M)\vec{Q}_{12}(12) | \hat{t}(E-E_R-\hbar^2 Q_{12}^2/2M) | \vec{k}_{n_0} - (m/M)\vec{Q}_{12}(12) \rangle, \quad (29)$$

where the notation is obvious.

This is our final result and I would like to discuss it in more detail. We see that TME is equal to the momentum distribution of the bound particle and to the two nucleon transition operator for the relative motion of the two protons. This two nucleon transition operator is completely off the energy shell as can be easily verified. This form for the momentum distribution for the bound particle and for the transition matrix element is also found when one replaces the distorted waves by their plane-wave counterparts except that in our case the momenta do not correspond to the true momenta but to the momenta in the plane-wave expansion. It should be pointed out that if the momenta which appear in the transition matrix element were the true momenta the matrix element would be half off the energy shell. Thus the use of distorted waves does not yield a single momentum distribution for the ejected particle but instead yields a group of such terms. In addition the transition matrix element has been completely placed off shell. The extraction of a momentum distribution may then be difficult. Of course this is a function of a great many things including energy.

In order to show the momentum distribution in a plane-wave expansion of a distorted wave we show the results of calculations for the distorted waves which would appear in the reaction $^{12}C(p,2p)^{11}B$ at an incident energy of 155 MeV. The reaction was assumed to be a co-planar one in which the outgoing protons had equal energies and were detected at equal angles. The results for the incident wave are shown in Fig. 2. Here we show the magnitudes of the expansion coefficients for orbital angular momentum zero and nine. Even for this relatively high energy we see that for $\ell = 0$ there is some spread in the coefficients. By the time we reach $\ell = 9$ it essentially looks like a pure plane wave. The optical parameters used in this calculation were obtained from Ref. 7. Figure 3 shows the corresponding results for the outgoing protons. Here, because of the decrease in energy there is even more of a spread in the coefficients and clearly shows there may be some problem in extracting momentum distributions or in using a factorized DWIA. The optical model parameters for this calculation were obtained from Ref. 8.

The calculation of the true matrix element TME will be very difficult, so let us look at some approximations. An effective t-matrix approximation was introduced by McCarthy and coworkers.[9] In this approach the transition operator is approximated by an effective operator t_{eff} such that

$$<\vec{r}\vec{R}|t_{eff}|\vec{r}'\vec{R}'> = \delta(\vec{R}-\vec{R}')\delta(\vec{r}-\vec{r}')t_{eff}(r), \quad (30)$$

where $\vec{R}$ is the center of mass coordinate of the two protons and $\vec{r}$ is their relative coordinate. Notice that this approximation neglects the center of mass motion completely and also makes t_{eff} local.

If we use this approximation in our original expression for the transition matrix element it is quite easy to show that

$$\text{TME} = <\vec{k}_{n_1}(12)|t_{eff}|\vec{k}_{n_0}(12)><\vec{k}_{n_1}+\vec{k}_{n_2}|u_{nj\ell m_\ell}|\vec{k}_{n_0}>. \quad (31)$$

Usually what one does is to empirically determine some $t_{eff}(r)$ from the analysis of nucleon-nucleon scattering. Of course the $t_{eff}(r)$ is completely determined on shell so that any off shell effects are hidden. The use of this approximation has met with only limited success and is undoubtedly not the answer.

Let us now reexamine our two expressions for TME. These are given by Eqs.(29) and (31). We observe that if we let $m/M \to 0$ in Eq.(29), i.e. neglect the center of mass motion of the two outgoing protons that we obtain for the exact expression

$$\text{TME} = <\vec{k}_{n_1}(12)|\hat{t}(E-E_R)|\vec{k}_{n_0}><\vec{k}_{n_1}+\vec{k}_{n_2}|u_{nj\ell m_\ell}|\vec{k}_{n_0}>, \quad (32)$$

where

$$\hat{t}(E-E_R) = v_{12} + \hat{t}(E-E_R)(E^{(+)}-E_R-K_{rel})^{-1}v_{12}. \quad (33)$$

We see that this expression looks much like the effective t-operator

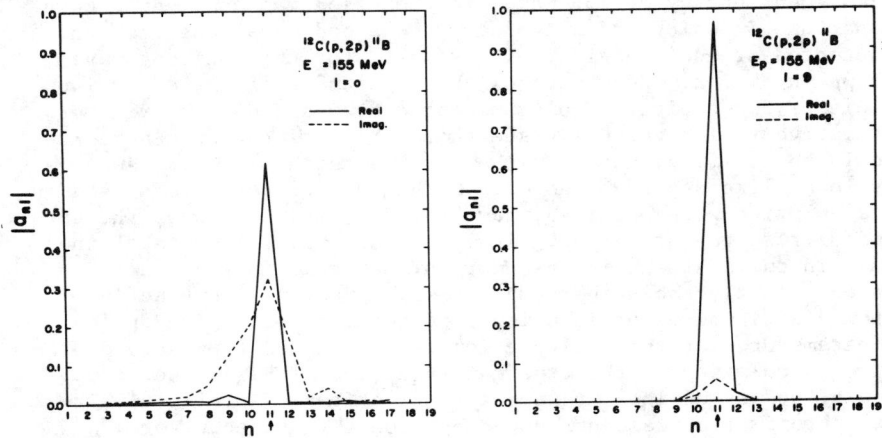

Fig. 2. The absolute magnitude of the expansion coefficients for 155 MeV proton incident on ^{12}C. (a) $\ell=0$. (b) $\ell=9$. The solid line shows the real part of the coefficient and the dashed line the imaginary part.

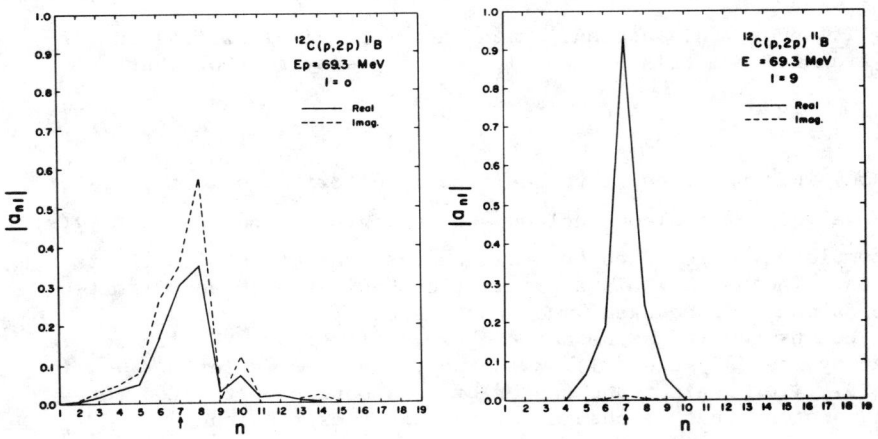

Fig. 3. The absolute magnitude of the expansion coefficients for 69.3 MeV protons incident on ^{11}B. See the caption to Fig. 2 for the description of the various quantities.

approach but uses the exact transition operator so that we can use its off shell properties. This also allows us to calculate an effective transition operator which may be useful in the study of nucleon inelastic scattering where the effective operator approach is used.

Thus, we have three methods for the calculation of (p,2p) reactions. We have developed a computer code AAB to perform such calculations. At present we only allow for the effective t-operator approach. We are currently working on the extension of the code to make use of the approximation given by Eq.(32). We have developed the code TOFF[9] to calculate the off shell matrix elements for most of the so called realistic nucleon-nucleon potentials.

Let me now present the results of a calculation for the reaction ^{14}N(p,2p)^{13}C (g.s.) for an incident energy of 46 MeV. For the effective interaction we used the one proposed by Picklesimer and Walker.[10] We used only that part of their interaction which depended upon the radial coordinate. The effective interaction can be written as

$$t(r) = t_R(r) + i\, t_I(r), \qquad (34)$$

where

$$t_j(r) = -t_o[a_{1j}\exp(-\mu_{ij}r)/\mu_{ij}r + a_{2j}\exp(-\mu_{2j}r)/\mu_{2j}r]. \quad (35)$$

The various parameters in Eq.(33) are shown in Table I.

Table I Values of parameters used in the effective interaction

t_o = 83 MeV	$\mu_{1R} = \mu_{1I}$ = 4 fm^{-1}		$\mu_{2R} = \mu_{2I}$ = 6 fm^{-1}
a_{1R} = 130	a_{1I} = 116	a_{2R} = -372	a_{2I} = -321

The results of the calculation for the angular correlation for the coplanar, equal energy sharing and equal angle case are shown in Fig. 4. The optical model parameters used in the calculation were obtained from the work of Watson et al.[11] The upper portion of the curve shows the comparison between the data[12] and the calculation which is a sum of contributions from a $1p_{3/2}$ and $1p_{1/2}$ proton knockout. The spectroscopic factors were obtained from the work by Cohen and Kurath.[13] The lower curve just shows the $1p_{3/2}$ contribution. One can see that it is small and can be neglected. The fit to the data is not too good; however, the magnitude is correctly predicted and some of the details are reproduced. However, as was mentioned previously one should not expect good agreement for the effective t-operator approach, particularly at the energies we have here. Other effective operators were tried and these gave essentially to the same results.

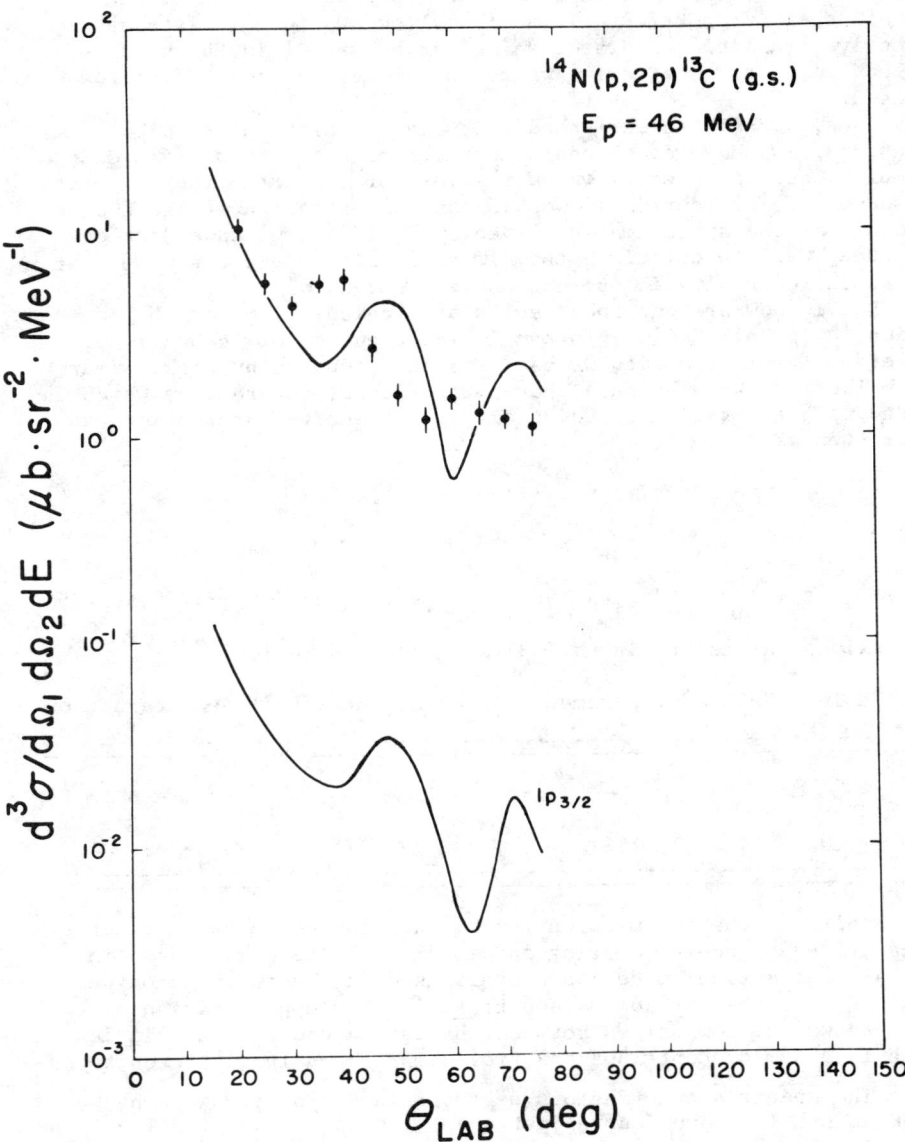

Fig. 4. The comparison of experiment and calculation for the reaction $^{14}N(p,2p)^{13}C(g.s.)$ at E_p = 46 MeV. The upper calculated curve is for both $p_{3/2}$ and $p_{1/2}$ proton removal. The lower curve is only $p_{3/2}$ removal. The data are taken from Ref. 12.

Figure 5 and 6 show the expansion coefficients obtained for this reaction. One can immediately see that the spread in moments is even larger here than in the previous case we discussed. Because of this large range of momentum contributions off the energy shell effects are probably important for the description of this reaction.

Our future studies will primarily be concerned with the study of the interplay between distortion and off the energy shell effects so that once the importance of these is understood we may be able to understand the nuclear structure. One thing in particular that we wish to study is the calculation of the optical potential from the elementary interactions. We can write the optical potential in momentum space as

$$\langle \vec{k}|U_{oo}^{(o)}|\vec{k}'\rangle = (N-1)\langle \vec{k}|\tau_{oo}|\vec{k}'\rangle$$
$$= (N-1)\langle \vec{k}\Phi_T|\tau|\vec{k}'\Phi_T\rangle, \qquad (36)$$

where

$$\tau = v_{12} + \tau A(E^{(+)} - K_1 - H_T)^{-1} v_{12}. \qquad (37)$$

H_T is the target Hamiltonian and A is an antisymmetrization operator. This transition operator is different from the one given in Eq.(4) which we use in our knockout calculations because of the appearance of the interaction of the bound particle with the residual nucleus by means of the interaction V_2. If we make the impulse approximation and neglect this term we can set $\tau = t$ and calculate the optical potential

$$\langle \vec{k}|U_{oo}^{(o)}|\vec{k}'\rangle = (N-1)\langle \vec{k}\Phi_T|t|\vec{k}'\Phi_T\rangle. \qquad (38)$$

If we then use this optical potential in our knockout calculations we then can do a consistent calculation, i.e. the distorted waves and the transition are obtained in a similar fashion. With our two nucleon transition matrix code TOFF mentioned earlier we can and will do these calculations.

REFERENCES

1. P A. Kazaks and R. D. Koshel, Phys. Rev. C1, 1906 (1970).
2. I. E. McCarthy, presented at this conference.
3. P. Nagel and R. D. Koshel, Phys. Rev. C13, 907 (1976).
4. P. Nagel and R. D. Koshel, Bull. Am. Phys. Soc. 18, 1400 (1973).
5. D. Robson and R. D. Koshel, Phys. Rev. C6, 1125 (1972).
6. P. Griffin, P. Nagel, and R. D. Koshel, J. Math. Phys. 15, 1913 (1974).
7. V. Comparat, R. Frascaria, N. Marty, M. Morlet, and A. Willis, Nucl. Phys. A221, 403 (1974).
8. C. B. Fullmer, J. B. Ball, A. Scott, and M. L. Whiten, Phys. Rev. 181, 1565 (1969).

9. R. D. Koshel and P. Griffin, Bull. Am. Phys. Soc. $\underline{20}$, 667, (1975).
10. A. Picklesimer and G. E. Walker, Bull. Am. Phys. Soc. $\underline{20}$, 690 (1975).
11. B. A. Watson, P. P. Singh, and R. E. Segel, Phys. Rev. $\underline{182}$, 977 (1969).
12. L. C. Welch, C. C. Chang, H. H. Forster, C. C. Kim, D. W. Devins, and P. A. Deutchman, Nucl. Phys. $\underline{A158}$, 644 (1970).
13. S. Cohen and D. Kurath, Nucl. Phys. $\underline{A101}$, 1 (1967).

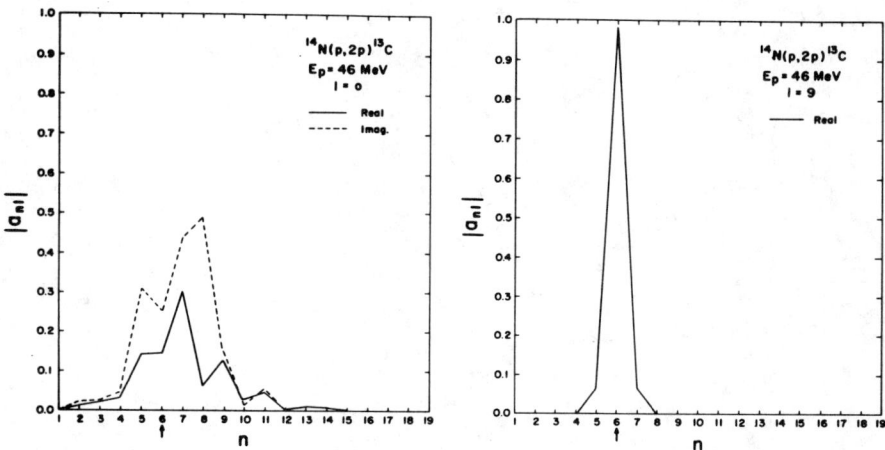

Fig. 5. The absolute magnitude of the expansion coefficients for 46 MeV protons incident on ^{14}N. See the caption to Fig. 2 for the description of the various quantities.

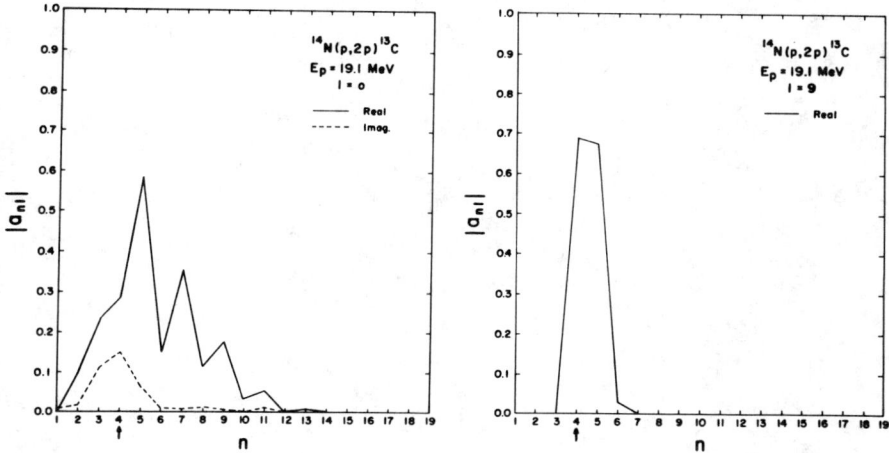

Fig. 6. The absolute magnitude of the expansion coefficients for 19.1 MeV protons incident on ^{13}C. See the caption to Fig. 2 for the description of the various quantities.

RECENT H.E.E.I.S. RESULTS ON THE COMPTON DEFECT*

A.D. Barlas, W. Rueckner and H.F. Wellenstein
Department of Physics, Brandeis University
Waltham, Massachusetts 02154

Electron Compton scattering experiments, previously described by Wellenstein and Bonham,[1,2] are performed by crossing a beam of high energy, but non-relativistic, electrons with a beam of atoms or molecules and measuring the energy loss spectrum over a range of scattering angles.

During the last three years a high energy electron impact spectroscopy (H.E.E.I.S.) apparatus has been constructed to undertake high precision Compton scattering experiments. Because Helium is the simplest experimentally feasible system for which the Compton profile has been calculated to better than 0.1% precision at the maximum, the energy loss spectra of helium were studied to check the performance of this new apparatus, the method of data analysis, and the theory used to convert cross sections to Compton profiles. It soon became apparent that the energy loss spectra taken over a range of scattering angles do not reduce by means of the binary encounter theory[2] (impulse approximation) to Compton profiles in agreement with theory. This disagreement is most evident in a shift of the experimental Compton peak from the peak predicted by the binary encounter theory. These results, as well as those obtained for molecular hydrogen, molecular deuterium, and molecular nitrogen are presented and discussed below. The most important experimental parameters are summarized in Table 1. A more detailed description of the apparatus is given elsewhere.[3]

The electron scattering cross section in the first Born approximation is given by

$$\frac{d^2\sigma_e}{dEd\Omega} = \frac{k_f}{k} \left(\frac{d\sigma}{d\Omega}\right)_R \left(\frac{K^2}{E}\right) \frac{df(K,E)}{dE} \qquad (1)$$

where k and k_f are initial and final momenta of the incident electrons, K is the momentum transfer, $(d\sigma/d\Omega)_R$ is the Rutherford cross section, and $df(K,E)/dE$ is the continuum generalized oscillator strength (GOS) defined by

$$\frac{df(K,E)}{dE} = \sum_n E_n |<\psi_n| \sum_{i=1}^{M} e^{i\vec{K}\cdot\vec{r}_i} |\psi_0>|^2 \delta(E-E_n)/K^2 \qquad (2)$$

To account for electron exchange, the Rutherford cross section in Eq. 1 must be replaced by the Mott cross section

$$\left(\frac{d\sigma}{d\Omega}\right)_M = F_{ex} \left(\frac{d\sigma}{d\Omega}\right)_R \quad \text{where} \quad F_{ex} = [1 - \frac{K^2}{K_s^2} + \frac{K^4}{K_s^4}]$$

Equation 1 is the motivation and starting point of the experiment.

Table 1 - Experimental Parameters

Electron Beam:

	Energy	35 keV
	FWHM	.2 to .4 mm
	Divergence	10^{-3} rad

Gas Beam:

	Density	.01 to .3 torr
	FWHM	15 mm

Energy Analyzer:

	Max. Res.	.25 eV
	Typical Res.	1 to 10 eV
	Acceptance Angle	5×10^{-3} deg.
	Energy Range	0 to 5 keV

Detector:

	Type	Surface Barrier
	Efficiency	approx. 60%
	Noise Threshold	20 to 30 KeV (Room Temp)
		13 keV at 9°C

Angular Measurement:

	Range	0-120°
	Working Range	1-20°
	Accuracy	± .002°
	Precision	± .001°

Using Eq. 1, the relative measured scattering intensities are converted to relative GOS which are then placed on an absolute scale by use of the Bethe sum rule discussed by Inokuti.[4]

$$\int dE \frac{df(K,E)}{dE} = N = \text{no. of target electrons}$$

where the integral is carried out over the entire energy loss spectrum. This is valid as long as the Born approximation is satisfied. With the GOS placed on an absolute scale and Binary Encounter (B.E.) and exchange corrections having been made, the connection between x-ray and electron scattering Compton Profiles, within the B.E.[3,2] theory, is given by:

$$\frac{df(K,E)}{dE} \simeq \frac{EJ(q)}{2K^3} = \frac{2\pi E}{2K^3} \int_{|q|}^{\infty} dp\, p\, \rho(p) \tag{3}$$

$$\text{with} \quad q = \frac{E-K^2}{2K} \tag{4}$$

Since all experiments measure the GOS as a function of energy loss for a fixed value of the scattering angle, the restriction that $df(K,E)/dE$ be integrated over a constant value of K for the Bethe sum rule to be valid poses a problem. This is not a serious problem, however, as K varies only 1 part in 10^4 over the region of interest. Our measurements on He and H_2 reported here exhibit deviations in that maxima of the observed inelastic peaks are shifted towards smaller momentum transfer. To account for this discrepancy, Inokuti has offered the following explanation. The exact expression for the GOS is given by

$$\frac{df(K,E)}{dE} = \frac{E/R}{2\pi\hbar(Ka_0)^2} \int_{-\infty}^{+\infty} dt\, \exp(-iEt/\hbar)\, \times$$

$$\times \left\{ \sum_j \sum_K \left< \exp\left[i\vec{K}\cdot(\vec{r}_j - \vec{r}_K) \right] \right> \times \exp\left[\frac{i}{\hbar} \int_0^t dt' \left(\frac{(\hbar K)^2}{2m} + \hbar \vec{K} \cdot \vec{p}_K(t') \right) \right] \right>$$

$$- \left| \sum_j \left< \exp(i\vec{K}\cdot\vec{r}_j) \right> \right|^2 \right\} \tag{5}$$

When E is large and one introduces the Impulse Approximation (IA), the Fourier component of the quantity in the brace on the right hand side of Eq. 4 must be taken. For large E, the only significant term in the brace is the one which varies most slowly with t. Thus one may replace $p_k(t')$ by p_k in the t' integration. This equivalent to neglecting all but the first term in

$$p_K(t) = \exp(iHt/\hbar)\bar{p}_K \exp(-iHt/\hbar)$$

$$= \bar{p}_K + \frac{(it)}{\hbar}[H,\bar{p}_K] + \frac{1}{2}\left(\frac{it}{\hbar}\right)^2\left[H,[H,\bar{p}_K]\right] + \ldots$$

Hence, integrating Eq. 4 with respect to t and t' yields

$$\frac{df(K,E)}{dE} = \frac{E/R}{(Ka_0)^2} \sum_j \sum_K \langle \exp[i\bar{K}\cdot(\bar{r}_j-\bar{r}_K)]\delta\left(\frac{(\hbar K)^2}{2m} + \frac{\hbar\bar{K}\cdot\bar{p}_K}{m} - E\right)\rangle \quad (6)$$

The argument of the δ-function may be rewritten as

$$\frac{(\hbar K)^2}{2m} + \frac{\hbar\bar{K}\cdot\bar{p}_K}{m} - E = \frac{(\hbar K+\bar{p}_K)^2}{2m} - \frac{\bar{p}_K^2}{2m} - E$$

This says that the gain of the kinetic energy of the kth electron, which has momentum $\bar{p}_k$ before the collision and momentum $\bar{p}_k + \hbar\bar{K}$ after the collision, is equal to the energy, E, transferred from the incident particle. That is to say, the electron behaves as if it were free, precisely in accordance with the neglect of the forces acting upon it within the atom as represented by the term

$$[H,\bar{p}_K] = i\hbar\bar{\nabla}_K u$$

The δ-function in Eq. 5 also implies that df(K,E)/dE is appreciable only for large $(\hbar K)^2/2m$, so long as E is large. Then $\exp[i\bar{K}\cdot(\bar{r}_j-\bar{r}_k)]$ is rapidly oscillating function of $\bar{r}_j-\bar{r}_k$ and makes no significant contribution unless j=k. This implies that electron correlations are unimportant for collisions with large momentum transfers. Therefore one may safely neglect the term with j=k to obtain the standard expression for the Binary Encounter Theory:

$$\frac{df(K,E)}{dE} = \frac{E/R}{(Ka_0)^2} \sum_j \langle \delta\left(\frac{(\hbar K)^2}{2m} + \frac{\hbar\bar{K}\cdot\bar{p}_i}{m} - E\right)\rangle \quad (7)$$

If we assume that ignoring the force term gives rise to the shift, one can pursue this argument with an elementary calculation suggested by R.J. Weiss.[5] Consider the He atom and let

E_0 = energy of the incident electron

$E_1 = E \sin^2\theta$, the energy of the ejected electron

θ = scattering angle

τ = collision time

d = distance ejected electron travels during collision time

Making use of the Heissenberg uncertainty principle, we have that

$$\tau = \frac{h}{E_0} = \frac{h}{mv_0^2}$$

$$d = \tau v_1 \quad \text{but} \quad v_1^2 = \frac{2E_1}{m} = \frac{2E_0 \sin^2\theta}{m} = v_0^2 \sin^2\theta$$

Therefore

$$d = \frac{h \sin\theta}{mv_0} = \frac{h \sin\theta}{\sqrt{2mE_0}}$$

Now E_2, the energy transferred to the spectator electron, is simply given by the product of the Coulomb force and the distance

$$E_2 = \frac{e^2}{r_{12}^2} d$$

where r_{12} is the average electronic distance in the atom or molecule. Hence,

$$\frac{E_2}{E_1} = A(E_0^{3/2} \sin\theta)^{-1} \qquad (8)$$

where

$$A = \frac{e^2}{r_{12}^2} \cdot \frac{h}{\sqrt{2m}}$$

From Eq. 4 we have that

$$\Delta q = \frac{1}{2K} \Delta E .$$

Then Eq. 8 becomes

$$\frac{\Delta E}{E} = \frac{2}{K} \Delta q$$

The Compton defect, Δq, thus becomes

$$\Delta q = A(E_0^{3/2} \sin\theta)^{-1} K \qquad (9)$$

Experimental measurements are given in Table II and Fig. 1. Energy loss spectra of 35 keV incident electrons were obtained at scattering angles ranging from 2° to 14° (K = 1.8 to 12.5 a.u.) using He, H_2, and D_2 as a target. The inelastic cross sections were converted to Compton profiles by applying Eq. 1 and 3, as discussed in Ref. 3. The peak of the Compton profile was determined by locating the center of the top 80% of the profile. These values appear in Table II. At scattering angles of less than 4° (K <3.0 a.u.) it was necessary, due to the asymmetry of the distribution, to extrapolate the center of the Compton profile to the peak of the profile. In all cases at least two measurements were taken by rotating the electron gun to both sides of zero angle. This provided an excellent check of the precision of zero angle. The standard deviation of the defect (Δq) given in Table II reflects the uncertainty in zero angle (0.002°) and the random error due to statistical fluctuations.

Table II - Defect Measurement for 35 keV Incident Electrons

θ (deg)	K (a.u.)	E(q=0)[a] (eV)	Δq(He)[b,c] (a.u.)	Δq(H$_2$)[b,c] (a.u.)
1.5	1.3	24.6		+.027±.014 (2)
2.0	1.8	44		-.041±.005 (4)
2.5	2.3	67	+.007±.009 (2)	-.034±.004 (4)
3.0	2.7	96	-.018±.007 (2)	-.028±.006 (4)
3.5	3.2	130		-.030±.003 (2)
4	3.6	176	-.028±.003 (4)	-.025±.003 (6)
5	4.5	273	-.027±.004 (4)	-.023±.002 (6)
6	5.4	394	-.028±.003 (4)	-.022±.002 (6) -.021±.003 (2)d
7	6.3	520		-.017±.002 (2)
8	7.1	681	-.019±.003 (6)	-.015±.003 (10) -.012±.003 (2)d
9	8.0	857		-.019±.004 (4)
10	8.9	1072	-.020±.003 (6)	-.014±.003 (4)
12	10.7	1524	-.019±.002 (6)	-.015±.008 (4)
14	12.4	2110	-.016±.003 (4)	-.012±.007 (4)

(a) energy loss at q=0
(b) equivalent energy defect, ΔE in eV, can be obtained by use of
 ΔE = (2E(q=0)/K)Δq
(c) no. of spectra taken in parenthesis
(d) defect measurement for D$_2$ at 6° and 8°

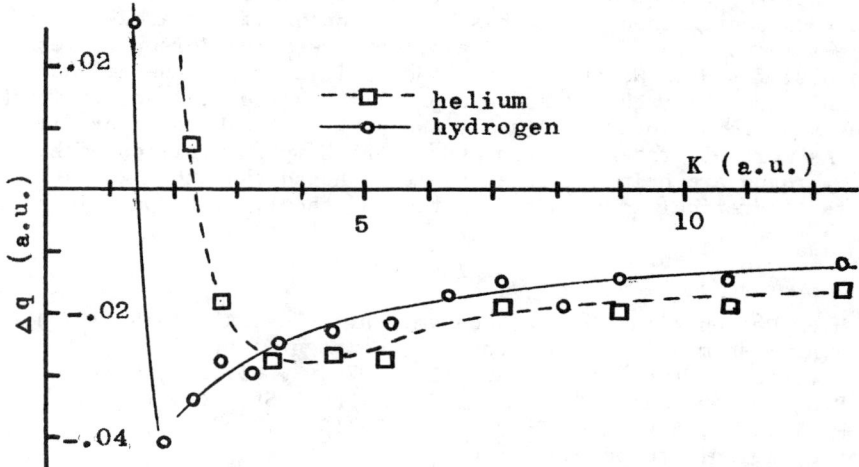

Figure 1. The momentum defect, the shift of the Compton peak from the binary encounter theory predicted position, is plotted versus momentum transfer for helium and molecular hydrogen.

Fig. 1 shows the defect Δq plotted versus momentum transfer. Eq. 9 closely follows the data if A is normalized to the data. This figure demonstrates the breakdown of the binary encounter approximation for K values of less than 15 a.u., with a maximum deviation of 2% for He at K=5 a.u. and more than 4% for H_2 at K=4 a.u. For helium the defect changes sign and becomes very large in the region of K less than 2 a.u. (1.5 a.u. for H_2). The binary encounter theory fails in this region as the binding energy of the target electron becomes comparable to the Compton shift. It should be noted that this occurs for He at about twice the K value compared to H_2, in good agreement with the ionization potential of the two systems. In the momentum transfer region from K = 5 to 12.4 a.u. the theoretical model (numerical values for He were used) in terms of the "spectator" electron is in good qualitative agreement with the He data represented. The factor of two discrepancy should not be surprising in lieu of the rather crude arguments made in the theoretical model. It is however most interesting to note that at large K values the defect for hydrogen is significantly less than the defect for helium, in qualitative agreement with the r_{12} values, the average distance between the two electrons, for these two systems. Finally it was found that the defect for H_2 and D_2 is the same within the experimental accuracy. This rules out the attempt to explain the defect in terms of the recoil of the atom or molecule. Preliminary measurements on N_2 have shown that the defect for N_2 at $\theta = 10°$ is $+2.0 \pm 4.0$ eV; this seems to rule out the possibility that the defect is due to the binding energy of the target electrons. A more detailed study of the defect in N_2 is currently undertaken.

This work has shown that the analysis of electron Compton profiles has to be done with great care due to the shortcoming of the binary encounter theory. It has however been demonstrated that at large momentum transfer (K > 10 a.u.) the relative defect is less than 0.2% for H_2. He should have the smallest r_{12} value and hence its shift can be regarded as an upper limit. The 0.2% defect for H_2 gives an error of 0.3% at $J(q=0)$, the peak of the Compton profile.

As more accurate measurements of the Compton profiles of helium and hydrogen are being undertaken, it is hoped that this work will incite theoretical work on the scattering theory of comparable accuracy.

REFERENCES

1. H.F. Wellenstein and R.A. Bonham, Phys. Rev. A$\underline{7}$, 1568 (1973).
2. R.A. Bonham and H.F. Wellenstein, Compton Scattering, Ch. 8: Electron Scattering, McGraw-Hill, 1976, B. William, Editor.
3. R.A. Bonham and C. Tavard, J. Chem. Phys. $\underline{59}$, 4691 (1973).
4. M. Inokuti, Rev. Mod. Phys. $\underline{43}$, 297 (1971).
5. R.J. Weiss, (to be published).

*This research was sponsored by the Research Corporation (RC 7769).

BORN APPROXIMATION CALCULATIONS OF THE COMPTON DEFECT IN ALUMINUM[+]

L. B. Mendelsohn
New School of Liberal Arts of Brooklyn College
Brooklyn, N.Y. 11210

H. Grossman[*]
Polytechnic Institute of New York, Brooklyn, N.Y. 11201

ABSTRACT

Excited State Hartree-Slater (EHS) Calculations have been performed on the systematics of Compton defects with increasing photon energy. For s states, the defects are typically negative and move toward zero with increasing momentum transfer. For p states the defects are typically positive. Thus both negative and positive defects for the entire atom are possible depending on the number of filled outer s and p orbitals and the experimental conditions. For the aluminum valence electrons, EHS calculations give a local maximum negative Compton defect of -2.5 ev. The experiment of R. Weiss gives a center of gravity Compton defect of -10 ± 7 ev.

INTRODUCTION

R. H. Weiss[1] has found for Mo K_β X-rays, scattered through 157.6° from an aluminum single crystal, that the center of gravity of its Compton profile was shifted about 1% towards the unmodified line. Noting that the Compton wavelength shift for free electrons is given by

$$\lambda_2^0 - \lambda_1 = (h/m_0 c)(1 - \cos \theta) \qquad (1)$$

where θ is the scattering angle, λ_1 is the incident wavelength, and λ_2^0 is the wavelength scattered from a free electron, this means that shift associated with the true center of gravity of the aluminum profile $(\lambda_2 - \lambda_1)$ is approximately 1% less than the shift given by Eq(1). Thus λ_2 is less than λ_2^0 and E_2, the energy of the scattered photon associated with the center of gravity, is greater than E_2^0, that is the photon on scattering loses less engergy than expected. This is referred to as a negative Compton defect. In the above experiment, Weiss found a defect of -10 ± 7 ev associated with the valence electrons. He also found sharp discontinuities in the slope at the Fermi momentum, both yielding a similar defect as the center of gravity. We also note that in Compton scattering it is convenient to use a momentum variable q where to a good approximation, $q \propto (\lambda_2 - \lambda_2^0)$. Thus a negative energy

---

+ Supported by ONR
* Supported in part by a CUNY Faculty Research Award

defect, as in this case, corresponds to a profile maximum at a negative value of q rather than at q = o for the free electron result. To our knowledge, this the first experiment to report a measurable defect since several experiments reported such shifts in the 1930's. In particular, Ross and Kirkpatrick[2] reported 1% to 2% negative defects for x-rays scattered from carbon and beryllium. DuMond and Kirkpatrick[3] observed a 1.25% negative defect for Mo K_α x-rays scattering from helium. Hughes and Mann[4] using 1000-4000 ev incident electrons observed positive defects on the resultant Compton profile, but saw no consistent pattern emerge. There was certainly some doubt of the validity of Ross and Kirkpatrick's results, especially since subsequent experiments by Kappeler in lithium using similar experimental techniques gave anomalous results[5]. Also P. Eisenberger[6] in 1970 using much more sophisticated equipment scattered MoK_α x-rays through 170° from helium and found that shifts of the order of the binding energy were clearly not indicated by his studies.

Therefore I find it most exciting to learn at this symposium that Weiss[7] now has new preliminary x-ray results for scattering from lithium and beryllium which also show negative defects of about 10 ev, and Barlas, Brueckner and Wellenstein[8] have precise results from high energy electron impact spectroscopy measurements for 35 kev incident electrons on helium which show negative defects ranging up to about 6 ev at all but the smallest scattering angle. In addition McCarthy et al[9] have observed positive 5 to 10 ev shifts (defects) of the quasi-free peak in the (e,2e) reaction for scattering of 200-800 ev electrons from He, Ar and Xe.

THEORETICAL CALCULATIONS

In the impulse approximation, the Compton profile $J(q)$ is given by

$$J(q) = \frac{1}{2} \int_q^\infty \frac{1}{p} |\chi(p)|^2 d^3p \qquad (2)$$

where $|\chi(p)|^2$ is the momentum probability distribution in the atom. It is quite clear that the impulse $J(q)$ is a symmetric monotonic decreasing function of q with increasing q and therefore can never exhibit any Compton defect. The maximum of $J(q)$ always occurs at $q = 0$. However first Born calculations of the Compton profile which explicitly include binding and a more accurate representation of the ejected electron than a plane wave (impulse) have been performed by Eisenberger and Platzman[10] and Mendelsohn and Biggs[11] for the hydrogenic 1s state. Such calculations which include electron binding effects lead explicitly to Compton defects. In the paper by Mendelsohn and Biggs[11], it is demonstrated that hydrogenic defects in energy and q for K-shell electrons are typically negative. It is also observed that for increasing values of momentum transfer, the defect approaches 0. In additon, following along the lines of F. Bloch[12], making the approximations $E_1/moc^2 < 1$ and [binding energy / (momentum transfer)^2] < 1, it is noted by these authors that a 1s Compton energy defect which is negative and equal to 1/6 of the binding energy is obtained.

First Born hydrogenic L-shell results have been given in a series of papers by Mendelsohn and Bloch[13,14,15]. In reference 13 it was found that profile calculations for MoK$_\alpha$ and AgK$_\alpha$ x-rays scattering from the L-shell of neon exhibit a positive defect in q of + 0.1. This was not observed experimentally but this is not surprising since x-ray Compton experimentalists usually use the maximum profile value to define q = 0 so as to be consistent with impulse. A summary discussion of the exact hydrogenic method can be found in the review by Mendelsohn and Smith[16].

More recently Grossman and Mendelsohn[17] have performed first Born calculations of Compton profiles using bound state Continuum Hartree-Slater wave functions in the frozen core approximation. Hartree-Slater bound slate and continuum wave-functions have been previously used to calculate photoelectric cross sections and very good agreement with the experimental results was obtained. For a typical Compton profile problem, many partial waves must be included in the continuum wave function to obtain convergance of the results to 4 figures. We have studied the systematics of Compton profiles associated with the orbitals of the K, L, M and N shells of Krypton. We have found that all s states typically exhibit negative defects which go toward q=0 with increasing momentum transfer. For p states, the typical defects are positive and these shift to more positive values of q with increasing momentum transfer. For d states the behavior of the profile itself and the defect is more complex than for either s states or p states. In general the maximum of the Compton profile occurs as a partial wave resonance effect. For s states, this is related to the first maximum in the ℓth continuum partial wave coinciding with the first maximum of the ℓth spherical Bessel function, these two functions entering into the integral which gives the s-state profile result.

From the above we see that either positive or negative shifts can be expected from our analysis. For an outer filled p shell, as is the case for the rare gases, we may expect the positive defects associated with the contribution of 6 p states to outweigh the negative defects associated with the 2 s states. This was the case in neon and Krypton, and may explain the positive (e,2e) results cited earlier. For He we would theoretically predict negative defects in accordance with the observations. Certainly an extensive comparison of theory with experiments of Barlas[8] is called for in the helium case and we shall begin such calculations shortly. Since Eisenberger[6] was looking for defects in helium of the order of the binding energy, rather than about 1/6 of the binding energy according to our results, this may account for his not observing such effects.

THEORETICAL DEFECT RESULTS FOR ALUMINUM

Since Weiss' measurement of the Compton defect in aluminum is the only recent experimental result for us to compare to, we have performed EHS profile calculations for this case and the results are given in Table I. In these calculations we have treated the valence electrons in aluminimum as if they were in atomic Hartree-Slater $3s$ (2 electrons) and $3p$ (1 electron) states with binding energies of 10.1 ev and 4.9 ev respectively. Since the work function of aluminum is about 2.3 ev, our atomic calculations would appear to overestimate the binding of the electron to the solid. Still the agreement with experiment as to the sign of the defect and its magnitude appear to be quite good. Clearly from the table the EHS profile is asymmetric, with the negative q side lying higher than the plus q side, in agreement with the experimental results. The magnitude of the EHS defect is -2.5 ev which occurs at $q = -.01$ on this grid. This represents the value of q where the valence electron profile has its maximum. We have previously tested the accuracy of our partial wave procedures on hydrogenic L-shell states, where an exact analytic profile result is known, and have observed that maximum errors of about ± 3 in the fourth figure may be generated by our numerical techniques. This would lead us to believe that uncertainties of about $\pm .01$ in q or about ± 2.5 ev should be associated with our defect results. As a first test of EHS defect predictions in solids, our aluminum results seem quite promising. A true test of the accuracy of the EHS method for calculating defects will come from an analysis of experiments performed on atomic gases.

TABLES

Table I Compton Profile and Defect in Aluminum for MO K_β X-Rays Scattering through 157.6°

q	E_2(ev)	J(q) $(3s)^2$	J(q) 3p	J(q) $(3s)^2 3p$
-.04	18261.1	2.2784	.9440	3.222
-.03	18258.5	2.2812	.9481	3.229
-.02	18256.0	2.2815*	.9522	3.234
-.01	18253.4	2.2794	.9562	3.236**
0.00	18250.9	2.2750	.9600	3.235
+.01	18248.4	2.2679	.9636	3.232
+.02	18245.9	2.2586	.9671	3.226
+.03	18243.3	2.2469	.9703	3.217
+.04	18240.8	2.2329	.9733+	3.206

+ The 3p state of aluminum has its maximum profile value of .9813 at q = +.09 corresponding to a positive evergy defect of 22.7 ev.

* The maximum profile value for 3s electrons occurs at q = -.02, which corresponds to a negative energy defect of -5.1 ev.

** The total valence electron profile has a maximum at q = -.01 corresponding to a negative energy defect of -2.5 ev.

REFERENCES

1. R.J. Weiss, Philos Mag $\underline{32}$, 247 (1975).
2. P.A. Ross and H. Kirkpatrick, Phys. Rev. $\underline{46}$, 668 (1934).
3. J.W. DuMond and H. Kirkpatrick, Phys. Rev. $\underline{52}$, 420 (1937).
4. A.L. Hughes and M.M. Mann, Phys. Rev. $\underline{53}$, 50 (1938).
5. M. Cooper, Adv. Phys. $\underline{20}$, 453 (1971).
6. P. Eisenberger, Phys. Rev. $\underline{A2}$, 1678 (1970).
7. R.J. Weiss, private communication.
8. A.D. Barlas, W. Rueckner and H.F. Wellenstein, private communication.
9. I.E. McCarthy, C.J. Noble, A. Ugabe and E. Weigold, private communication.
10. P. Eisenberger and P.M. Platzman, Phys. Rev. $\underline{A2}$, 415 (1970).
11. L.B. Mendelsohn and F. Biggs, in Inner Shell Ionization Phenomena and Future Applications, R.W. Fink editor, 1142 (1973).
12. F. Bloch, Phys. Rev $\underline{46}$, 674 (1934).
13. L.B. Mendelsohn, B.J. Bloch and V.H. Smith Jr, Phys.Rev. Lett. $\underline{31}$, 266 (1973).
14. B.J. Bloch and L.B. Mendelsohn, Phys. Rev. $\underline{A9}$, 129 (1974).
15. L.B. Mendelsohn and B.J. Bloch, Phys. Rev. $\underline{A\,12}$, 551 (1975).
16. L.B. Mendelsohn and V.H. Smith Jr, Chapter 6 in COMPTON SCATTERING, McGraw Hill (1976).
17. H. Grossman and L.B. Mendelsohn, to be published.

THE COMPTON DEFECT: IS THERE A SHIFT IN THE COMPTON PEAK?

I. E. McCarthy[+]
R. A. Bonham[*]
Indiana University, Bloomington, Ind. 47401

ABSTRACT

Recently observed shifts in the position of the maximum of Compton profiles using x-ray and HEEIS techniques and in the origin of momentum distributions using the (e,2e) reaction are discussed. The observed shifts for He from the free electron peak position are qualitatively explained by use of a distorted wave formalism. The problems remaining in the development of a quantitative theory for peak shifts are discussed.

INTRODUCTION

For some years one of the most successful probes for properties of electron wave functions in atoms and molecules has been the Compton profile, given for a closed-shell system by

$$J(q) = 2\pi \Sigma_i \int_{|q|}^{\infty} p\,dp\, \rho_i(p), \qquad (1)$$

where $\rho_i(p)$ is the momentum density of the ith electron in the independent particle (Hartree-Fock) model.

The Compton profile is measured by high energy inelastic scattering of photons or electrons (High Energy Electron Impact Spectroscopy, HEEIS, or the electron Compton effect). It is assumed that, at the high energy (several tens of kilovolts) of the experiment, the binding energy of the electrons may be neglected and the Born approximation can be used to calculate the matrix element M_K for the basic knockout reaction, $(\gamma, \gamma'e)$ for the Compton effect for HEEIS.

The kinematic variables relevant to the reaction are $\vec{k}_0$, E_0, momentum and kinetic energy of the incident particle, $\vec{k}_f$, E_f and $\vec{k}$, E', the corresponding quantities for the faster of the two outboing particles and the unobserved electron respectively, $\vec{q}$, the recoil momentum of the residual ion (its energy may be neglected) and I, the ionization energy of the ejected electron. We consider one electron at a time and keep in mind the case of helium, which is a simple illustration. The variables $\vec{k}_0$, $\vec{k}_f$ and E_f are measured in the experiment.

[+]This author wishes to acknowledge research support by Indiana University. Permanent address is School of Physical Sciences, The Flinders University of S. Australia, Bedford Park, South Australia.

[*]This author wishes to acknowledge support by the National Science Foundation, Grant No. GP-41983X and to the Donors of the Petroleum Research Fund, administered by the American Chemical Society.

Making the binary encounter and impulse approximations, the cross section for the reaction is abbreviated as follows.

$$\sigma_C = (\frac{d\sigma}{d\Omega})_{free} \, f_K \, J(p), \qquad (2)$$

where $(d\sigma/d\Omega)_{free}$ is the free scattering cross section of the relevant two bodies calculated at some appropriate energy and f_K is a kinematic factor appropriate to the reaction.

From the beginning of the study of the Compton effect,[1] it has been known that a more rigorous description includes the ionization energy I. In fact the cross section for the reaction is strictly defined as the integral over the direction of the momentum of the unmeasured electron (its energy is known) of the cross section σ for the kinematically-complete knockout process $(\gamma,\gamma'e)$ or $(e,2e)$.

In the Born approximation σ is a function only of q, where

$$\vec{q} = \vec{K} - \vec{k},$$
$$\vec{K} = \vec{k}_0 - \vec{k}_f, \qquad (3)$$
$$E = E_0 - E_f.$$

Normally the angle θ between $\vec{k}_f$ and $\vec{k}_0$ is kept fixed and the energy difference E is the experimental variable.

Defining x as the cosine of the angle between the momentum $\vec{k}$ of the ejected electron and the measured momentum transfer $\vec{K}$, we have

$$\sigma_C \propto \int_{-1}^{1} dx \, \sigma(q). \qquad (4)$$

In the plane wave Born approximation $\sigma(q)$ is proportional to $|M|^2$ where

$$M = \int d^3r \, \exp[i(\vec{k}_0 - \vec{k}_f - \vec{k}) \cdot \vec{r}]\psi(\vec{r})$$

and

$$|M|^2 = \rho(q). \qquad (5)$$

For s-state electrons $\rho(q)$ is peaked at $q = 0$, the Compton peak.

The limits q_1 and q_2 of the integration in (4) are given by

$$q_1^2 = (K-k)^2, \quad q_2^2 = (K+k)^2. \qquad (6)$$

Using Rydberg atomic units, the conservation of energy is expressed by

$$E_0 = E_f + E' + I,$$

where, in the photon case

$$k_0^2 = E_0^2/4c^2, \quad k_f^2 = E_f^2/4c^2, \quad k^2 = (E-I) + (E-I)^2/4c^2 \qquad (7)$$

and, in the electron case

$$k_0{}^2 = E_0 + E_0{}^2/4c^2, \quad k_f{}^2 = E_f + E_f{}^2/4c^2, \quad k^2 = (E-I) + (E-I)^2/4c^2 \quad (8)$$

For nonrelativistic ejected electrons the lower limit of integration in Eq. 4 is given by

$$q_1 = (E_r - I)^{1/2} - K, \quad (9)$$

Where E_r is the relativistically-corrected energy difference. The upper limit is essentially infinite in normal experimental conditions.

The Compton peak, $q = 0$ occurs, according to (9), not at the free electron value $E_r = K^2$ but at $E_r = K^2 + I$. We call this a positive shift of I relative to the free electron value $E = K^2$.

Calculation of the "exact Born approximation", first given by Bethe[2] in 1930, as described in Inokuti[3] for atomic hydrogen, in which the ejected electron is represented by a Coulomb wave give the opposite result. The Compton peak has a negative shift but it is equal to only about one sixth of the ionization energy.

We are thus forced to conclude that any <u>first Born</u> treatment for either the electron or x-ray case with a plane wave function for the ejected electron will predict Compton profile peak positions shifted toward <u>higher</u> energy loss by an amount equal to the binding energy or some appreciable fraction thereof. Further, the situation in (e,2e) reaction experiments will have a q scale given by Eq. 9 in which the peak position will be shifted from the free electron value, $K = \sqrt{E_r}$, by $+ I/2\sqrt{E_r}$. How does this view compare with current experimental observations?

Because of recent technological advances all experimental techniques discussed so far in this note are capable of making measurements with accuracy greater than the predicted shifts which are of the order of magnitude of the ionization potentials I. As pointed out in the accompanying notes[4-6] observed shifts from x-ray HEEIS experiments are in the <u>negative</u> direction from the peak positions expected on the basis of the free electron theory but the (e,2e) results are observed to have a <u>positive</u> shift. The magnitude of these shifts are about 3/4 to 1/4 of the first ionization potential ranging from -10 eV for the x-ray cases cited by Weiss[4] to -4.5 eV for HEEIS measurements on He cited by Wellenstein.[5] The (e,2e) results exhibit a + 9 eV shift.

The x-ray and HEEIS values represent 20 - 30 eV negative shifts from the predicted first Born plane wave results which include binding effects. The situation is even more serious when one considers the case of N_2 where nearly 30% of the electrons are bound by more than 400 eV. HEEIS experiments show no shift in the Compton profile peak for core electrons greater than 10 eV.[7] The simple theory (Eq. 9) would predict a shift of + 400 eV.

We thus have a paradox. Correct inclusion of the ionization energy in Eq. 9 produces shifts of the Compton peak that are totally in disagreement with experiment, by orders of magnitude in many cases.

The paradox is approximately resolved as follows. The Born approximation[12,13] may be used to describe elastic scattering at the relevant energies, because the average potential felt by the unbound particles is small compared to their incident energies. However we are concerned here with absolute energy shifts and in the electron scattering cases, though presumably not for the incident and scattered photon wave vectors in the x-ray case, we must take the distortion of the continuum waves by the core of the system into account. Another way of saying this is that coupling of the inelastic Compton channels to the elastic channel cannot be neglected, at least in (e,2e) experiments. This is not so surprising as it might appear when one considers recent studies on higher Born corrections to inelastic scattering cross sections.[9-11] More importantly note that in systems containing two or more electrons distorted waves must be used to properly describe the ejected electron wave function in all three experiments.

It has been shown in calculating (e,2e) cross sections[8] that the distorted-wave off-shell impulse approximation is correct within experimental error. In this approximation $(d\sigma/d\Omega)_{free}$ is replaced by the correct half-off-shell t-matrix element, which in the present calculation still may be taken outside the $\vec{k}$ or $\vec{q}$ integration. The factor M becomes

$$M = \langle \chi^{(-)}(\vec{k}_f) \chi^{(-)}(\vec{k}) | \chi_0^{(+)}(\vec{k}_0) \Psi(I) \rangle, \quad (10)$$

where the $\chi^{(\pm)}$ are the wave functions for elastic scattering of the appropriate energy in the appropriate two-body subsystem and $\Psi(I)$ is the overlap between the wave function for the target and the wave function for the final ion state.

Furthermore, it has been shown[12,13] that an adequate description of the distorted waves is given at energies greater than a few hundred volts by the averaged eikonal approximation

$$\chi^{(+)}(\vec{k},\vec{r}) = \exp(-\gamma kR) \exp(i\vec{\mathcal{K}}\cdot\vec{r}) \quad (11)$$

where $e^{-\gamma kR}$ is a normalizing factor and in which the electron wave number k is replaced by an averaged complex wave number $\mathcal{K}$:

$$\mathcal{K} = k' + i\gamma k = (1 + \bar{V}/2E + i\bar{W}/2E)k. \quad (12)$$

Here $\bar{V}$ and $\bar{W}$ are the average real and imaginary potentials felt by the electron at the collision point. The imaginary potential does not produce any significant peak shifts. This argument is similar to using a WKB approximation for the electron wave functions in which the potential $V(r)$ in the wave vector $\sqrt{E-V(r)}$ is replaced by an average value. We consider replacing k by k' in Eq. (5). Note again that the (e,2e) experience suggests that the average eikonal wave description for the ejected wave is an adequate one.

Since the bound electron requires energy I to lift it out of its potential, a good first approximation to the average potential felt by an electron at the collision point is a value slightly greater than I, because the potential energy for a given orbital is equal to I at the classical turning points but is usually greater between. We therefore compute the matrix element (5) as an approximation to (10) using

$$k_\mu'^2 = E_\mu + I + (E_\mu + I)^2/4c^2, \qquad (13)$$

where μ labels each electron.

We have
$$|M|^2 = \rho(q') \qquad (14)$$
where
$$\vec{q}' = \vec{k}_0' - \vec{k}_f' - \vec{k}'. \qquad (15)$$

The external kinematics are of course unchanged so that the energy conservation laws (7) and (8) still hold, as does the momentum conservation law (3).

Considering the lower limit q_1 in Eq. 6 in terms of the new electron wave numbers k_μ' of (13), we find upon substitution of

$$\vec{k}_0' = \vec{k}_0(1 + \frac{\overline{V}}{2k_0^2})$$

$$\vec{k}_f' = \vec{k}_f(1 + \frac{\overline{V}'}{2k_f^2})$$

$$\vec{k}' = \vec{k}(1 + \frac{\overline{V}''}{2k^2})$$

into

$$\vec{q_1}' = \vec{k}_0' - \vec{k}_f' - \vec{k}' = \vec{K}' - \vec{k}'$$

that the numerator of

$$\frac{K' + k'}{K' + k'} q_1 = \frac{K'^2 - k'^2}{K' + k'}$$

can be written through terms to first order in the $\overline{V}$'s as

$$K^2 - E + I - 2(E-I)\frac{\overline{V}''}{2k^2} + (K^2+E)\frac{\overline{V}}{2k_0^2} + (K^2-E)\frac{\overline{V}'}{2k_f^2} \qquad (16)$$

The free electron peak position occurs at $K^2 = E$ so that the peak shift δ is given by

$$\delta = I - \frac{(E-I)\overline{V}''}{k^2} + \frac{E\overline{V}}{k_0^2}$$

We can now analyze the three different experimental types as follows. In the x-ray case we assume that no non Born effects occur and hence $\overline{V} = \overline{V}' = 0$. We assume the proper choice of $\overline{V}''$ is $I+\epsilon$ where I is the ionization potential of the ejected electron and ϵ is the increase of $\overline{V}$ over I. Since $K^2 = E-I$ the shift is negative and is simply given in first order by

$$\delta = -\epsilon$$

In the HEEIS experiment the incident and final energies differ by no more than about 5% and the ejected electron energy is of the same order of magnitude as in the x-ray case. We can assume that $\bar{V} \sim \bar{V}'$ and that $\bar{V}'' \sim I+\epsilon$ with ϵ the same as in the x-ray case. The shift is given by

$$\delta \cong -\epsilon + \frac{K^2 \bar{V}}{k_0^2}$$

where if $\bar{V}$ is of the order of I, $K^2 \sim 100$ a.u. and $k_0^2 \sim 25$ keV the shift can be expected to be about 1.3 eV less negative than in the x-ray case for targets with similar binding energies.

In (e,2e) experiments the scattered and ejected energies are identical and are approximately half the incident energy. In addition the ejected energy is of the same order of magnitude as in the x-ray and HEEIS cases. Hence $\bar{V}' = \bar{V}'' \sim I+\epsilon$ and the shift is given by

$$\delta = -\epsilon + \frac{1}{2}\bar{V}.$$

Because the incident energy is only twice the ejected energy we can expect $\bar{V}$ to be approximately the same or $I + \epsilon$ which leads to the result $\delta = 1/2(I-\epsilon)$ for which the sign of the shift is not certain.

Using Wellenstein's shift of -4.5 eV for He from HEEIS measurements leads to a prediction of + 10 eV for the (e,2e) shift which is in excellent agreement with the results of Weigold and McCarthy. Direct determination of $\bar{V}$ for the outermost s and p electrons in He, Ne, Kr and Xe by Giardini-Guidoni et al, at 400 eV to 2.6 keV yields values on the average less than I. This would infer negative values of ϵ in disagreement with the higher energy results. It must be emphasized, however, that the uncertainties involved in the determination of the $\bar{V}$ values for higher energy electrons are larger by virtue of their deeper probing of the atomic potential. In fact the data of Wellenstein et al., show that for larger impact parameters (smaller K) the shift ϵ changes sign. The x-ray results are at least in qualitative agreement with the ideas presented here.

It is important to summarize the remaining questions left unanswered by the present discussion.
1) How to best carry out rigorous calculations of the peak shifts, especially in the x-ray case?
2) What are the exact details of the mechanism for coupling to the elastic channel and is elastic channel coupling really important in the HEEIS case?
3) How to define the q scale?

Let us discuss these points further. The problem of peak shift calculations has been solved in the (e,2e) case by fitting eikonal waves to elastic scattering and by carrying out rigorous but very difficult calculations using more realistic models for the distorted waves.[8] In the HEEIS case these approaches become very difficult since one is looking for a shift in the wave vector of the order of

$I/2k_0^2$ which is around .01% at the usual HEEIS energies. This is further complicated by point 2) raised above since it is not known for certain whether the shift can be obtained within the framework of elastic static potential scattering or if polarization of the target (coupling to inelastic channels) must also be included or if the shift is completely explained by the distortion in the ejected wave as is presumably the case with x-rays. It may be possible to use a WKB approximation utilizing Hartree-Fock potentials in distorted wave Born calculations for atoms. The molecular case is clearly going to present a much more difficult problem.

The x-ray case bears special mention. The local momentum arguments can be applied to the ejected electron wave function in systems other than hydrogen. The hydrogen case should yield about a -2 eV peak shift unless the first Born application to photon is correctly chosen for this case. It can also be argued that the electron case is similar since the ejected (slow) electron sees the pure Coulomb field of the nucleus for the case of hydrogen. Note that for more than two electrons in the system and single ionization that the ejected electron wave function will not be a pure hydrogenic Coulomb function although the qualitative agreement of the exact hydrogenic results for H with the experimental findings in He would seem to suggest that the Coulomb wave approximation is a reasonable starting point.

Thus in the photon scattering case for systems containing two or more electrons we would expect the local momentum argument to yield a similar result to the electron scattering case independent of whether or not the Born approximation for photon scattering was valid. Eisenberger and Platzman[15] have investigated some higher order Born corrections in the photon case and have concluded that they were small. It might be worthwhile to look at this problem again in view of the smallness of the effects sought for and the approximations made by these authors in their analysis.[16]

The definition of a q scale is a difficult problem. We would suggest using both the relativistic definitions

$$q = \frac{E_r - K^2 + \epsilon}{2K} \text{ and } q_1 = \frac{E_r - K^2 + \epsilon}{\sqrt{E_r} + K}$$

with ϵ determined experimentally to see which of the two scales agrees best with experiment. A second alternative, clearly to be preferred by experimentalists, is to report the observed intensities on an energy loss scale rather than a q scale, which we have shown here depends critically on the scattering dynamics. This leaves the theorist with the task of performing the necessary distorted wave transformations on the theoretical momentum distributions in order to compare theory with experiment. This last alternative is the one currently in use in (e,2e) reaction studies. So far inclusion of more realistic distorted waves has always led to better agreement between experiment and theory in (e,2e) work.[8]

We have thus shown qualitatively that the large-shift paradox is unreal. Further, the simple model employed has been successful in correlating the results of three very different types of experiment. To compute small shifts we must use a better approximation than (13). At least in the electron scattering case we must bo beyond use of a plane wave description of the ejected electron and we may have to go beyond the first Born approximation. In view of the success of (10) in the (e,2e) reaction[8], calculations using this approximation with distorted waves $\chi_\mu^{(\pm)}$ related to elastic scattering will be performed in the near future.

REFERENCES

1. G. Wentzel, Z. Physik 58, 348 (1929).
2. H. Bethe, Ann. Physik 5, 325 (1930).
3. M. Inokuti, Rev. Mod. Phys. 43, 297 (1971).
4. R. J. Weiss, accompanying note and Phil. Mag. 32, 247 (1975).
5. H. F. Wellenstein, accompanying note.
6. E. Weigold and I. E. McCarthy, accompanying note.
7. T. C. Wong, J. S. Lee, H. F. Wellenstein and R. A. Bonham, Phys. Rev. A 12, 1846 (1975).
8. I. E. McCarthy and E. Weigold, Physics Reports (to be published).
9. S. Geltman and M. B. Hidalgo, J. Phys. B 4, 1299 (1971); B 5, 617 (1972).
10. W. Huo, J. Chem. Phys. 57, 4800 (1972).
11. R. A. Bonham, J. Elect. Spec. and Rel. Phenom. 3, 85 (1974).
12. E. Weigold, S. T. Hood and I. E. McCarthy, Phys. Rev. A 11, 566 (1975).
13. A. Ugbabe, E. Weigold and I. E. McCarthy, Phys. Rev. A 11, 576 (1975).
14. A. Giardini-Guidoni, R. Tiribelli, D. Vinciguerra, R. Camilloni, G. Stefani and G. Missoni, accompanying note.
15. P. Eisenberger and P. M. Platzman, Phys. Rev. A 2, 415 (1970).
16. It should be noted that the aim of Ref. 10 was to justify use of the authors' impulse approximation by comparison with exact hydrogenic calculations. Because their impulse approximation developed as a time expansion of the exact first Born expression it is clear that inclusion of corrections to all orders would have to yield the observed peak shift. Hence the authors were not concerned with energy shifts of the order of 13.6 eV in their work but only the fact that the two calculations agreed within such energy scale shifts.

SPECIAL ASPECTS OF THE NUCLEAR PROBLEM

R.M. Thaler
Case-Western Reserve University, Cleveland, OH

Redish and Stephenson both said some things I might have said. Now, in critique, I am at liberty to recapitulate. Fortunately for us all, I missed Walker's talk; so I may in my ignorance cover some of the same ground. The style, however, is uniquely my own. I also missed Koshel's talk, but I want to ignore that because he is doing honest calculations and is, hence, most vulnerable to criticism.

Let us begin by discussing scattering, in the case where there are only two body forces. In that case we may write

$$T = \sum_i v_{oi} + \sum_i v_{oi} G_o T \quad . \tag{1}$$

This is an operator equation which requires a great deal of careful discussion, all of which we shall ignore for the purpose at hand. All I want to stress now is that if we find T, we know the scattering amplitude, etc.

We follow the Watson prescription and define T_i such that

$$T \equiv \sum_i T_i \tag{2}$$

and

$$T_i = v_{oi} + v_{oi} G_o \sum_j T_j \quad . \tag{3}$$

These two equations are obviously equivalent to Eq(1). Now, of course, the subscripts i and j refer to target particles. The projectile is labelled as zero. Thus the scattering from a single target constituent particle i is

$$t_{oi} = v_{oi} + v_{oi} G_o t_{oi} \quad , \tag{4}$$

from which we can immediately obtain

$$T_i = t_{oi} + t_{oi} G_o \sum_{j \neq i} T_j \quad . \tag{5}$$

Iteration of Eq (5) then leads to the solution of Eq.(1) in the form

$$T = \sum_i t_{oi} + \sum_i t_{oi} G_o \sum_{j \neq i} t_{oj} + \sum_i t_{oi} G_o \sum_{j \neq i} t_{oj} G_o \sum_{k \neq j} t_{ok}$$
$$+ \sum_i t_{oi} G_o \sum_{j \neq i} t_{oj} G_o \sum_{k \neq j} t_{ok} G_o \sum_{\ell \neq k} t_{o\ell} + \ldots \quad . \quad (6)$$

Eq. (6) is the Watson multiple scattering series, in which the first term is the so-called single scattering part and the second term the double scattering piece and so on.

The Watson multiple scattering series can be rearranged as

$$T = \sum_i t_{oi} + \{\sum_i t_{oi} G_o \sum_{j \neq i} t_{oj} + \sum_i t_{oi} G_o \sum_{j \neq i} t_{oj} G_o t_{oi} + \ldots\}$$
$$+ \{\sum_i t_{oi} G_o \sum_{j \neq i} t_{oj} G_o \sum_{k \neq i,j} t_{ok} + \ldots\} + \ldots \quad . \quad (7)$$

Identification of the terms in Eq. (7) so grouped allows us to present Eq. (7) as

$$T = \sum_i t_{oi} + \sum_{i<j} (t_{o,ij} - t_{oi} - t_{oj}) + \ldots \quad , \quad (8)$$

where

$$t_{o,ij} = (v_{oi} + v_{oj}) + (v_{oi} + v_{oj}) G_o\, t_{o,ij} \quad , \quad (9)$$

or

$$t_{o,ij} = (t_{oi} + t_{oj}) + t_{oi} G_o t_{oj} + t_{oi} G_o t_{oj} G_o t_{oi} + \ldots$$
$$+ t_{oj} G_o t_{oi} + t_{oj} G_o t_{oi} G_o t_{oj} + \ldots (10)$$

Eq. (8) is the correlation expansion of Ernst, Londergan, Miller and Thaler. The first term in this expansion represents the scattering from pairs of particles ($t_{o,ij}$) corrected, of course, for that part of the scattering from the pairs which has already been accounted for in the first term. The series so generated is obviously a finite series with no more terms than there are target particles.

The point of this becomes apparent when we calculate the matrix elements appropriate to elastic scattering. In that case we see immediately that the first term in Eq. (8), the single scattering term, is weighted by the single particle density of the target. The second term in Eq. (8), the pair term, is weighted by the "pair" density, and so on. Thus if we wish to learn about the pair distribution of the target as well as the single particle distribution we must be in a position to identify the contributions of the first two terms in this correlation expansion to real data. The second term is, of course, the part which will give us information about correlations in the target.

However, it is not so easy as that. In nuclear physics, we do not have a fundamental theory which tells us unambiguously that the nuclear forces are two-body forces. For atomic and molecular physics the situation is much simplified by the fact that we know the force. To see how complicated matters can become without such a clarifying postulate, let us look at the equivalent potential argument of nuclear physics. Let us suppose that there exists a Hamiltonian H which describes the interaction of one nucleon with another. Then we can define the two-body interaction potential to be

$$V = H - p^2/2m . \quad (11)$$

We may then examine the result of a unitary transformation U upon the Hamiltonian H, i.e. we take

$$\hat{H} \equiv UHU^+ . \quad (12)$$

If the transformation is finite ranged, i.e. if $U \to 1$ for large distances, then clearly all the energy eigenfuctions of $\hat{H}$ become identical with those of H at large distances. Thus the bound state eigen energies and the continuum energy phase shifts are identical. Therefore, if we define a potential $\hat{V}$ as

$$\begin{aligned}\hat{V} &= \hat{H} - p^2/2m \\ &= UHU^+ - p^2/2m \\ &= UVU^+ + U(p^2/2m)U^+ - p^2/2m,\end{aligned} \quad (13)$$

we have obviously defined a two-body potential $\hat{V}$ which is equivalent to V in that it will produce the same discrete spectrum and the same scattering results. In the technical parlance of scattering theory, we say that

$$\hat{t}_{oi} = t_{oi} , \quad (14)$$

on the energy-shell. However, if we calculate $\hat{t}_{o,ij}$, that is, if we calculate the scattering from a pair of particles i and j with the two force given by $\hat{V}$, we find that

$$\hat{t}_{o,ij} \neq t_{o,ij} , \quad (15)$$

neither on-shell nor off-shell. However, we must be able to distinguish $\hat{t}_{o,ij}$ from $t_{o,ij}$ in order to use scattering experiments as a source of information regarding the pair correlations.

To understand more clearly what is happening here, we may wish to study the entire many-body system. We postulate a Hamiltonian for the many-body system under discussion to be H_{MB}, where

$$H_{MB} = K + V, \qquad (16)$$

with K taken to be the many-body kinetic energy operator and V taken to be the many-body potential energy operator. As before, then, with the help of a finite-ranged many-body unitary operator U, we may find

$$\hat{V}_{MB} = UHU^+ - K \qquad (17)$$

$$= \hat{V}_2 + \hat{V}_3 + \ldots$$

$$= \sum_{\alpha\beta} \hat{v}_{\alpha\beta} + \sum_{\alpha>\beta>\gamma} w_{\alpha\beta\gamma} + \ldots \qquad (18)$$

The expansion of Eq.(18) is a cluster expansion. We note that all physical results obtained with $\hat{V}_{MB}$ are necessarily identical to those obtained with V. However, to obtain these identical results we must include the three-body forces w, etc. The two-body forces $\hat{v}$ of Eq.(18) are just the equivalent two-body forces generated in Eq. (13). Thus so long as we are ignorant as to the existence and/or nature of the three-body forces we are at an impasse. The expansion of Eq.(8) unlike that of Eq.(6) remains unaltered if there are many-body forces. Hence it should be clear that we can study correlations if we know t_{oi} and $t_{o,ij}$. However, we must know $t_{o,ij}$. We cannot calculate $t_{o,ij}$ from t_{oi} or equivalently from two-body forces. Our conclusion is that one cannot distinguish a cluster from a correlation and that we must know the forces unambiguously to determine correlations.

We could take the attitude that nucleon-nucleon data could give us information about the two-body part of the interaction $v_{\alpha\beta}$. With this information we could calculate the deuteron wave function. From nucleon-deuteron scattering data we might determine $w_{\alpha\beta\gamma}$, and with this information calculate ^{3}He and ^{3}H. The data on nucleon-^{3}He and D-D scattering would then yield information on the four body force, and so on. If nuclear physics as so described were a "closed" system, this might be a reasonable program. We might then seek a many-body transformation such that

$$V_{BEST} = \Lambda V \Lambda^+ + \Lambda K \Lambda^+ - K$$

$$= \text{Two body only.} \qquad (19)$$

In general this is not possible, however. In which case we might seek that transformation which made the cluster series of Eq. (18) most rapidly "convergent". This is inverted from the atomic (moleculer, solid state...) problem. There we assume that we <u>know</u> the electromagnetic forces, and hence in principle could calculate <u>all</u> the relevant wave functions. Practical considerations lead to the use of the electron, e.g., as a probe to determine densities. We can do this because we <u>know</u> the force unambiguously, and thus can unravel densities.

Nuclear physics, on the other hand, has a duality of purpose. We must find the forces and we must find the wavefunctions. We should like to do this unambiguously and use the one set of information as a cross-check on the other. This is impossible in a closed system such as we are presently discussing. In the early 1930's Wigner proposed a systematic development of nuclear theory based on the assumption of local two-body forces. Under such a hypothesis, the two-body force can be determined from two-body data. Thereafter nuclear physics would proceed in a manner analagous to atomic physics.

Nuclear physics is not closed. Rather we have an embarrassment of riches in the strong, electromagnetic and weak interactions, in neutrinos, photons, electrons, muons, pions, kaons, nucleons,... and their antiparticles. Let's look at an example of how the interplay of the different forces and particles can alter our closed argument. We shall take nucleon-nucleon scattering as our illustrative example. Let us accept as necessary the hypothesis that the nuclear force is charge independent. Then we could search out a charge-symmetric potential V such that for proton-proton scattering

$$V_{pp} = V + V_c , \qquad (20)$$

where V_c is the Coulomb potential, and for neutron-neutron scattering

$$V_{nn} = V . \qquad (21)$$

We can then generate an equivalent interaction $\hat{V}_{pp}$ for proton-proton scattering as

$$\hat{V} = UHU^+ - \frac{p^2}{2m} - V_c$$

$$= UVU^+ + U \frac{p^2}{2m} U^+ - \frac{p^2}{2m} + UV_c U^+ - V_c . \qquad (22)$$

Similarly, the same transformation can yield the equivalent interaction for neutron-neutron scattering to be

$$\hat{W} = UVU^+ + U \frac{p^2}{2m} U^+ - \frac{p^2}{2m} . \qquad (23)$$

Thus

$$\hat{V} - \hat{W} = UV_c U^+ - V_c \quad , \tag{24}$$

which clearly indicates that this condition is uniquely satisfied by only one of the family of "equivalent" interactions. This argument is a simplified version of one advanced by Peter Sauer. We note that our knowledge of the electromagnetic interaction was the key to this discussion, that allowed us to remove some of the ambiguity necessarily inherent in our treatment of nuclear physics as a "closed" system.

The electron as a probe gives us the clearest information on distributions, because we presumably understand the electromagnetic interaction. But: (1) we obtain the charge distribution, not the matter distribution from such data; (2) the radiative corrections are difficult and very large. We have reason to believe that these corrections are handled correctly, but there is a great deal of theory embedded in those experimental results. There are definite questions as to the meaning of the form factors we extract from electron scattering at large momentum transfer. However, it must be emphasized that distributions so obtained are not subject to the theoretical uncertainties of the "equivalent" interaction.

There is a kind of complementarity in nuclear physics. First, we may begin with the idea that we know the interaction between the probe (projectile) and the nucleon. With this knowledge we can use the projectile to determine nuclear properties. On the other hand, we may begin with a knowledge of nuclear properties. In that case the nucleus may serve as a "filter" to study differing aspects of the probe-nucleon interaction.

An example of this is that above $\sim$ 100-200 MeV in many circumstances the deuteron is simply as close as we can get to a neutron target. This is especially important in certain two-body interactions which we may wish to study in which neither particle may readily serve as a target. We can study n-p scattering by scattering neutrons from hydrogen, but if we are interested in the kaon interaction with the neutron we must use the neutrons in nuclei as the target, because we can make neither neutron nor kaon targets. So we necessarily do both things. We must use nuclei to learn about the kaon. We can also use the kaon to probe the nucleus.

Let us consider another example. Can we use the pion as a probe of nuclei? The pion is the mediator of the strong force. Baryon number is conserved, but meson number is not. The pion is like the photon in that way. This means that two-body information is likely to be less complete than nucleon-nucleon two-body information. We may describe the π-N interaction in terms of a field theory. The basic graph is

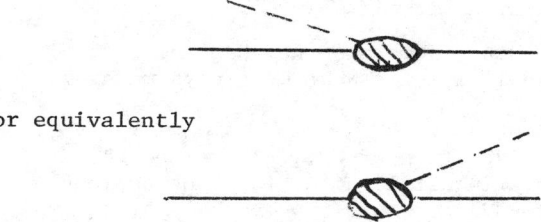

or equivalently

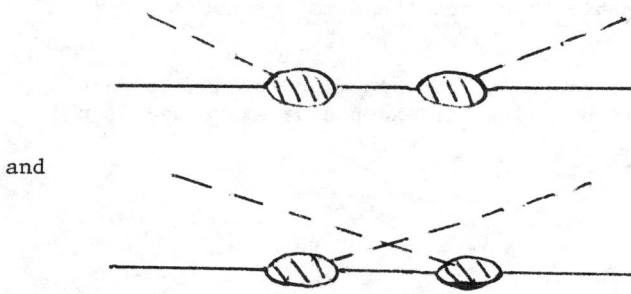

The first picture indicates a pion incident upon a nucleon and absorbed. The second picture represents a pion created at the nucleon site. The blob represents the nucleon form factor or extended vertex. We need to find this vertex function from experiment because we cannot calculate it. To do so we may wish to study π-N scattering, since the processes indicated above (pion absorption by a free nucleon) is kinematically forbidden.

For π-N scattering the relevant graphs are

and

The pion-nucleon t-matrix is given by iterates of the above diagrams. So we may hope to find the extended vertex by means of the study of pion-nucleon elastic scattering. But we must note that a new element enters our considerations of many-body scatterings. The absorption of a pion by a pair of nucleons, as in the diagram below

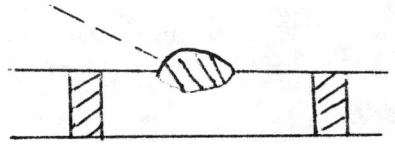

is possible. Thus we observe that something new happens in the presence of the second nucleon. This is what we always mean by a three body force.

If we now return to consideration of Eq. (8), we see that this development of the scattering still holds for pion-nucleus scattering. However the term

$$T_{3-body} = \sum_{i<j} \{t_{o,ij} - t_{oi} - t_{oj}\} \qquad (25)$$

may be more important in general for pion scattering than for nucleon scattering. Furthermore, we cannot calculate $t_{o,ij}$ from Eq. (10), but rather we must study some three-body events in addition to the two-body data.

Here then the complementarity becomes very apparent. Complex nuclear data are absolutely essential for study of pionic properties, on the one hand. On the other hand, the pion represents an exceptionally valuable probe for the study of nuclear correlations.

A number of **other** processes come to mind in this connection. Pion charge exchange scattering for example as expressed in the diagrams

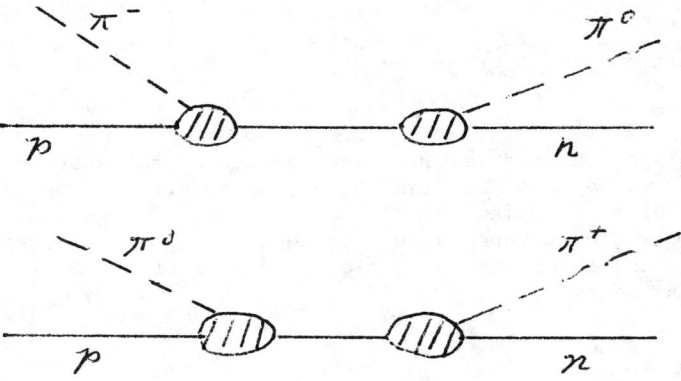

and double charge exchange

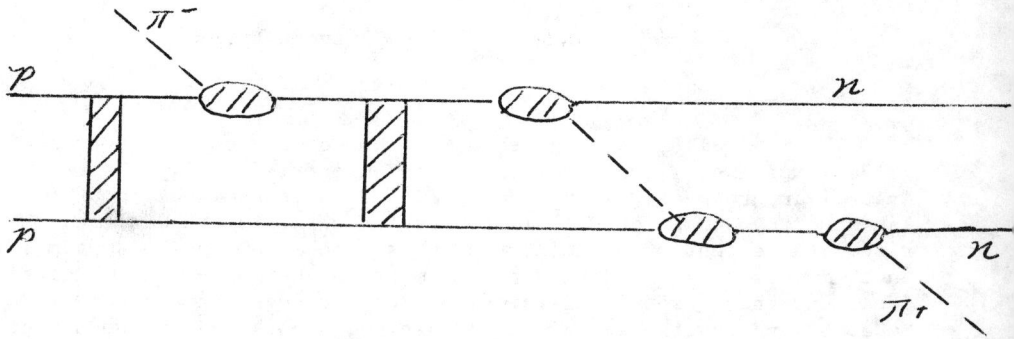

The purpose of this remark is to indicate that the same basic building blocks enter into a variety of processes in rather different ways, so that it is not entirely hopeless to expect that we can relate one set of data to another and let a consistent picture emerge.

WORKSHOP SUMMARY

R.A. Bonham

Chemistry Dept., Indiana University

The impetus for holding this workshop came from the success of (e,2e) experiments on atoms and molecules carried out by the Australian group at Flinders University under the direction of Ian McCarthy and Erich Weigold. These experiments are designed to reveal the extent of one electron orbital behavior in the system under investigation and have proved most successful.[1] The original ideas for these experiments, interestingly enough, came from McCarthy and coworkers[2] and Neudatchen and coworkers[3] in the Soviet Union. Both groups specialized in nuclear theory and the (e,2e) idea had in fact been an outgrowth of earlier nuclear (p,2p) experiments which had been judged less than completely successful. It was felt in some circles that proton beams of energy higher than those employed in the first experiments would be required before the potential of the (p,2p) method could be realized. An obvious question arises as to whether or not the newly gained successful experiences in (e,2e) could in some way provide new insights into the older (p,2p) experiments.

Because new (p,2p) experiments had been planned for the Indiana University Cyclotron Facility (IUCF) as a joint U.S.-Australian collaboration sponsored by the National Science Foundation (NSF) and because there was substantial interest in charge and momentum density measurements on atoms and molecules in the chemistry department at Indiana University it seemed natural to propose a Workshop/Seminar to be held at Indiana University. The most novel aspect of the proposal was bringing together two apparently very disparate groups of researchers. Could these two groups communicate with each other? More importantly, could either group profit from the experiences of the other?

During the winter of 1975 Ian McCarthy was brought to Indiana University for two months as a visiting Professor. During this period the advanced planning for the workshop was carried out. The agenda was designed to outline and review the theory and experiments behind investigations of the basic reactions in atomic-molecular work, (e,e'), (e,2e), positron anhilitation and (γ,γ') and those in nuclear studies such as (p,2p), (e,e') and (e,e'p), for observing single particle behavior. This process was to occupy the first two days. After the introductory period discussions of recent results would take place and the conference would close with a summary and analysis of accomplishments.

Presentations on the two subject areas were interspersed with the hope that this would prevent fragmentation of the workshop into discussions involving only sub groups of specialists.

The workshop opened on May 31, 1976 and ran to June 4, 1976. McCarthy from Flinders was the lead off speaker and reviewed the theoretical ideas behind (e,2e) and (p,2p) reaction studies. Examples of experimental results from both reactions were presented. M. Inokuti from Argonne National Laboratory reviewed the Born-Bethe theory of electron scattering and its application to obtaining structural information on atoms and molecules. G. Walker from Indiana reviewed nuclear knockout reaction theory. This was followed by a presentation on experimental results for (e,e') studies at low incident electron energies (< 1 keV) by E.N. Lassettre from Carnegie-Mellon University. The closing talk of the day was given by P. Roos of the University of Maryland who surveyed previous work on the (p,2p) reaction.

This alternating format continued through the conference's second day. R.A. Bonham, Indiana, presented results for atomic-molecular (e,2e') studies at high incident energy (25 keV). This was followed by G. Igo, UCLA. who discussed information available from nuclear knockout studies. P. Berko, Brandeis, discussed the use of positron anihilation to obtain momentum distributions in atoms, molecules and solids. D. Devins, Indiana, described the I.U.-Melbourne (p,2p) collaborative efforts. This was followed by Erich Weigold, and Flinders, who presented the Australian (e,2e) results. P. Radvanyi, Orsay, presented some recent results from the French group. The Tuesday afternoon presentations were completed by E.F. Redish, Maryland, who discussed nucleon knockout reaction mechanisms. The evening was devoted to presentations of contributed papers. J.H. Moore, Maryland, discussed low energy secondary electron ejection from Ar and He. P. Winkler, Erlangen-Nurnberg, discussed ($\hbar\omega$,2e) reactions applied to atoms and molecules. D. Hutcheon, TRIUMF, presented the first (p,2p) data obtained using a polarized proton beam. V.H. Smith, Queens, discussed recent rigorous quantum mecahnical configuration interaction calculations on the momentum density and Compton profile of H_2. The evening was concluded by L.C. Snyder, Bell Labs, who discussed the effects of chemical bonding on Compton profiles.

Wednesday started off with a review of quantum chemical RPA computational techniques by G. Williams, Sydney. B.M. Spicer, Melbourne, then surveyed the present status of research in nuclear physics in Australia. This was followed by a panel discussion on theoretical and experimental problems in momentum distribution determinations in atoms, molecules and solids organized and presided over by R.J. Weiss, Watertown, with help from Weigold, Berko and Smith. G. Stephenson, Los Alamos, discussed off-shell effects in nucleon knockout reactions. A. Giardini-Guidoni,

Frascatti, discussed the current state of (e,2e) experiments in the Frascatti laboratory. R. Cole, U.S.C., then surveyed plans for (p,2p) reactions at LAMPF. This was followed by a discussion of (π,πn) reactions by Maris, Rio Grande do Sul. The three body problem and its relation to nuclear structure reactions was presented by I. Sloan, Univ. of New South Wales. The last talk of the day was given by N. Avery, CSIRO, Melbourne, on inelastic electron collisions in solids and plans to attempt (e,2e) experiments with solids.

On Thursday morning R.D. Koshel, Ohio University, discussed distorted wave calculations and his progress in testing the validity of factorization of the t-matrix. A mini symposium was then held on recent experimental observations of a Compton defect. E. Weigold, Flinders, presented (e,2e) results for a defect. R.J. Weiss, Watertown, presented x-ray scattering evidence for a defect and D. Barlas, Brandeis, presented similar results from (e,e') experiments. R.J. Weiss and I. McCarthy, Flinders, then presented two different model explanations of the defect. L. Mendelsohn, CCNY, presented current results on a priori calculations of the continuum generalized oscillator strength which showed promise of explaining the observed Compton defects. The closing presentation of the day was given by R.M. Thaler, Case Western Reserve, on his observations of the present state of nuclear structure theory. The afternoon was spent on organized tours of the Quantum Chemistry Program Exchange (QCPE), the Electron Impact Spectroscopy Laboratory located in the Chemistry Department and the Indiana University Cyclotron Facility. The evening was devoted to an informal banquet followed by discussion and planning session for future activities of the group. It was suggested that an application be made for a Gordon Conference on the experimental measurement of charge and momentum density. In addition it was proposed that another workshop be organized on knockout reactions to be held two years hence in Australia. An American and an Australian committee were selected for organizing this venture.

On Friday morning M. Inokuti and E.F. Redish collaborated to present an extremely interesting comparison of the two fields represented at the Conference. They then led an active group discussion on the presented materials. The workshop was officially adjourned by the week long discussion leader and chairman, I. McCarthy at 11 A.M. in order that transportation to flight departures could be arranged.

In summarizing the results of the conference perhaps the one thing which was most impressive was the degree to which the atomic-molecular research area has paralleled the nuclear structure area. This was manifested by extensive interactions during and after formal presentations between representatives of the two groups. The degree of communication that was possible seemed to surprise most people in attendance.

Ian McCarthy in his opening presentation pointed out that if dimensionless units based on the barn (10^{-24} cm^2) for the cross section and 10^{12} cm^{-1} for momentum transfer were used in the nuclear case and Rydberg atomic units in the atomic-molecular case, then the magnitudes of momentum transfer and cross sections were almost identical in the two areas. This scaling observation proved most helpful to the participants during the remainder of the conference.

It became apparent that both in the theory of (e,e') and (e,2e) studies and in nuclear reaction theory a central approximation is the factorization of the scattering amplitude into a half off shell t-matrix times a structure factor. The nuclear theorists appeared to be ahead of their atomic-molecular counterparts in going beyond this approximation. Walker, Redish, Koshel and others described more sophisticated calculations. Koshel described a mathematical technique, an expansion of a distorted wave in terms of plane waves, for testing the factorization approximation which showed promise of also being useful in the atomic molecular case.

Experimental evidence presented by Weigold appeared to indicate the possibility of a failure of the factorization approximation. It was pointed out that such a failure might also be expected in the nuclear case. Theory suggests that the factorization approximation might be expected to hold better in the symmetric non copalnor geometry for (e,2e) reactions than for the symmetric coplanor case. This is in agreement with Weigold's experiments. Redish has discovered theoretical reasons why the symmetric non coplanar (p,2p) reaction at a scattering angle of 45° may be advantageous. It was pointed out that at IUCF this geometry is difficult to attain experimentally.

Roos remarked after Weigold's presentation that he had seen more (e,2e) results in 30 minutes than (p,2p) results in his lifetime. This led to a discussion of the merits of having a dedicated facility for a single purpose. The (p,2p) people seemed to feel that a proposal for more time or a dedicated facility for (p,2p) work would be premature until additional experimentation was carried out to establish the usefulness of the reaction to elucidate nuclear structure. An example in the opposite direction occurred in Roos' presentation where he discussed (p,2p) experiments carried out in a fixed geometry with variable energy detection in order to test the kinematics of the reaction theory. Such tests are possible and highly desirable in (e,2e) work but had not been thought of.

In the closing summary Inokuti and Redish presented a point

by point comparison of the two fields. It was pointed out that
there is a one to one comparison between the two areas. The
incident projectile energy range where simple reaction theories
can hope to be applied is almost identical in the scaled units.
The (e,2e) case ceases to be easily interpretable when the
incident energy reaches the point where relativisitc quantum
electrodynamics must be employed to understand what is happening.
In the nuclear case the high energy limit is obtained when the
incident projectile energy is sufficient to produce mesons.

   The feeling of a majority of those attending the conference
was that it was a success. People appeared to be genuinely
surprised at how much they had in common with the "other" area
of research. Communication between the two groups took place
at a fairly sophisticated level and a number of useful new ideas
appeared to have been transmitted between the two sides in
addition to the expected exchanges within the individual groups.
The organizers and the attendees at the workshop appear to be
unanimous in proposing that a followup conference in two years
would be both desirable and highly beneficial to all concerned.

REFERENCES

1. S.T. Hood, I.E. McCarthy, P.J.O. Teubner and E. Weigold, Phys. Rev. $\underline{A8}$, 2494 (1973).

2. I.E. McCarthy, E.V. Jezak and A.J. Kromminga, Nuclear Physics $\underline{12}$, 274 (1959).

3. Z. Mathies and V.G. Neudatchin, JETP $\underline{45}$, 131 (1963).

WORKSHOP PARTICIPANTS

Avery, Neil	CSIRO, Tribophysics Univ. of Melbourne, AUSTRALIA
Barlas, Dilek	Brandeis University, Waltham, Mass.
Bent, R.	Physics Department, Indiana University
Berko, P.	Brandeis University, Waltham, Mass.
Bonham, R.A.	Chemistry Department, Indiana University
Brussell, M.K.	University of Illinois, Urbana, Ill.
Chant, N.S.	University of Maryland, College Park, Md.
Cole, R.	University of Southern Calif., Los Angeles, Calif.
Coplan, M.	University of Maryland, College Park, Md.
Devins, D.	Physics Department, Indiana University
Dillon, M.	Carnegie-Mellon University, Pittsburgh, Pa.
Emery, G.	Physics Department, Indiana University
Giardini-Guidoni, A.	Frascati Insitute, ITALY
Hagstrom, S.	Chemistry Department, Indiana University
Hennino, T.	Institut de Physique Nucleaire, Orsay, FRANCE
Hutcheon, D.	TRIUMF, British Columbia, CANADA
Igo, G.	University of Southern Calif., Los Angeles, Calif.
Iijima, T.	Chemistry Department, Indiana University
Inokuti, M.	Argonne National Laboratory, Argonne, Ill.
Kennerly, R.	Chemistry Department, Indiana University
Koshel, R.D.	Ohio University, Athens, Ohio
Lambert, J.	Georgetown University, Washington, D.C.
Lassettre, E.N.	Carnegie-Mellon University, Pittsburgh, Pa.
Le Bornec, Y.	Institut de Physique Nucleaire, Orsay, FRANCE
Lee, J.S.	Chemistry Department, Indiana University
McCarthy, Ian	Flinders University, Bedford Park, SOUTH AUSTRALIA
Malik, B.	Physics Department, Indiana University
Maris, Th.A.J.	Federal Univ. of Rio Grande do Sul, Porto Alegre, BRAZIL
Mendelsohn, L.	Brooklyn College of the City Univ. of N.Y., Brooklyn, N.Y.
Moore, J.H.	University of Maryland, College Park, Md.
Newton, R.	Physics Department, Indiana University
Ostlund, N.	Chemistry Department, Indiana University
Pollock, R.	Physics Department, Indiana University
Pugh, H.G.	University of Maryland, College Park, Md.
Radvanyi, P.	Institut de Physique Nucleaire, Orsay, FRANCE
Redish, E.F.	University of Maryland, College Park, Md.
Roos, P.	University of Maryland, College Park, Md.
Shull, H.	Chemistry Department, Indiana University
Sloan, I.H.	University of New South Wales, Kensington, AUSTRALIA

Smith, Vedene	Queen's University, Kingston, Ontario, CANADA
Snyder, Larry	Bell Telephone Labs, Murray Hill, N.J.
Spicer, B.M.	University of Melbourne, Melbourne, AUSTRALIA
Stephenson, G.J.	Los Alamos National Lab., Los Alamos, New Mexico
St. John, W.	Chemistry Department, Indiana University
Szabo, A.	Chemistry Department, Indiana University
Thaler, R.M.	Case Western Reserve University, Cleveland, Ohio
Tripathi, A.N.	University of Cincinnati, Cincinnati, Ohio
Walker, G.	Physics Department, Indiana University
Weigold, E.	Flinders University, Bedford Park, SOUTH AUSTRALIA
Weiss, R.J.	Army Materials and Mechanics Res. Center, Watertown, Mass.
Williams, G.	University of Sydney, Sydney, AUSTRALIA
Wills, J.	Physics Department, Indiana University
Winkler, P.	University of Erlangen-Nurnberg, GERMANY
Yates, A.	University of Cincinnati, Cincinnati, Ohio

AIP Conference Proceedings

		L. C. Number	ISBN
No. 1	Feedback and Dynamic Control of Plasmas (Princeton) 1970	70-141596	0-88318-100-2
No. 2	Particles and Fields - 1971 (Rochester)	71-184662	0-88318-101-0
No. 3	Thermal Expansion - 1971 (Corning)	72-76970	0-88318-102-9
No. 4	Superconductivity in d- and f-Band Metals (Rochester 1971)	74-18879	0-88318-103-7
No. 5	Magnetism and Magnetic Materials - 1971 (2 parts) (Chicago)	59-2468	0-88318-104-5
No. 6	Particle Physics (Irvine 1971)	72-81239	0-88318-105-3
No. 7	Exploring the History of Nuclear Physics (Brookline, 1967, 1969)	72-81883	0-88318-106-1
No. 8	Experimental Meson Spectroscopy - 1972 (Philadelphia)	72-88226	0-88318-107-X
No. 9	Cyclotrons - 1972 (Vancouver)	72-92798	0-88318-108-8
No. 10	Magnetism and Magnetic Materials - 1972 (2 parts) (Denver)	72-623469	0-88318-109-6
No. 11	Transport Phenomena - 1973 (Brown University Conference)	73-80682	0-88318-110-X
No. 12	Experiments on High Energy Particle Collisions - 1973 (Vanderbilt Conference)	73-81705	0-88318-111-8
No. 13	π-π Scattering - 1973 (Tallahassee Conference)	73-81704	0-88318-112-6
No. 14	Particles and Fields - 1973 APS/DPF Berkeley)	73-91923	0-88318-113-4
No. 15	High Energy Collisions - 1973 (Stony Brook)	73-92324	0-88318-114-2
No. 16	Causality and Physical Theories (Wayne State University, 1973)	73-93420	0-88318-115-0
No. 17	Thermal Expansion - 1973 (Lake of the Ozarks)	73-94415	0-88318-116-9
No. 18	Magnetism and Magnetic Materials 1973 (2 parts) (Boston)	59-2468	0-88318-117-7
No. 19	Physics and the Energy Problem - 1974 (APS Chicago)	73-94416	0-88318-118-5
No. 20	Tetrahedrally Bonded Amorphous Semiconductors (Yorktown Heights, 1974)	74-80145	0-88318-119-3

No. 21	Experimental Meson Spectroscopy - 1974 (Boston)	74-82628	0-88318-120-7
No. 22	Neutrinos - 1974 (Philadelphia)	74-82413	0-88318-121-5
No. 23	Particles and Fields - 1974 (APS/DPF Williamsburg)	74-27575	0-88318-122-3
No. 24	Magnetism and Magnetic Materials - 1974 (20th Annual Conference San Francisco)	75-2647	0-88318-123-1
No. 25	Efficient Use of Energy (The APS Studies on the Technical Aspects of the More Efficient Use of Energy)	75-18227	0-88318-124-X
No. 26	High-Energy Physics and Nuclear Structure - 1975 (Santa Fe and Los Alamos)	75-26411	0-88318-125-8
No. 27	Topics in Statistical Mechanics and Biophysics: A Memorial to Julius L. Jackson (Wayne State University-1975)	75-36309	0-88318-126-6
No. 28	Physics and Our World: A Symposium in Honor of Victor F. Weisskopf (M.I.T. 1974)	76-7207	0-88318-127-4
No. 29	Magnetism and Magnetic Materials - 1975 (21st Annual Conference, Philadelphia)	76-10931	0-88318-128-2
No. 30	Particle Searches and Discoveries - 1976 (Vanderbilt Conference)	76-19949	0-88318-129-0
No. 31	Structure and Excitations of Amorphous Solids (Williamsburg, Va., 1976)	76-22279	0-88318-130-4
No. 32	Materials Technology - 1975 (APS New York Meeting)	76-27967	0-88318-131-2
No. 33	Meson-Nuclear Physics - 1976 (Carnegie-Mellon Conference)	76-26811	0-88318-132-0
No. 34	Magnetism and Magnetic Materials - 1976 (Joint MMM-Intermag Conference, Pittsburgh)	76-47106	0-88318-133-9
No. 35	High Energy Physics with Polarized Beams and Targets (Argonne, 1976)	76-50181	0-88318-134-7
No. 36	Momentum Wave Functions - 1976 (Indiana University)	77-82145	0-88318-135-5